MATHEMATICS
Models
of the Real World

Peter Lancaster
Professor of Mathematics
The University of Calgary

PRENTICE-HALL, INC.
Englewood Cliffs, New Jersey

Library of Congress Cataloging in Publication Data

LANCASTER, PETER, mathematician.
　　Mathematics.

　　Bibliography:　p.
　　Includes index.
　　1. Mathematics—1961-　2. Mathematical models.
I. Title.
QA39.2.L3　　510　　75-17716
ISBN 0-13-564708-8

PRENTICE-HALL INTERNATIONAL, INC., *London*
PRENTICE-HALL OF AUSTRALIA, PTY. LTD., *Sydney*
PRENTICE-HALL OF CANADA, LTD., *Toronto*
PRENTICE-HALL OF INDIA PRIVATE LIMITED, *New Delhi*
PRENTICE-HALL OF JAPAN, INC., *Tokyo*
PRENTICE-HALL OF SOUTHEAST ASIA (PTE.) LTD., *Singapore*

To the three Jays

CONTENTS

PREFACE

In the last several years, pre-university mathematics education in North America has been transformed quite dramatically. A new emphasis on precision of language and concept has become apparent and, unfortunately, this has often been accompanied by a corresponding reduction of useful exercise in mathematics. Consequently, exposure to the demands of first-year mathematics courses at a university is often a traumatic experience.

There seem to be two basic problems here. First, the machinery required to make use of the axiomatic method is being taught without giving significant exercise in the axiomatic method itself. Second, the *usefulness* of mathematics in helping us to understand the world and its inhabitants is not sufficiently emphasized.

The second of these problems is our prime concern in this book. Nevertheless, in the context of a variety of applications, we also try to give some feeling for the limitations, as well as the capabilities, of mathematics as applied to our environment. Thus, we try to explain and discuss both the mathematics involved and the situation to which it is applied. This concern for precision in the use of mathematics has meant that the methods used have had to be very limited. In a one-term (39-hour) course, for which this book is designed, it is not possible to develop the methods of the calculus, for example, and also to apply them. It has therefore been an important factor in collecting material that every subject must be amenable to treatment, in the first instance, by the methods which we can be confident that high school students have met. This puts a serious limitation on the applications we are able to handle. With more mathematical techniques at our disposal, a wider variety of physical problems would be amenable to mathematical analysis.

Every reader should have some familiarity and facility with the following: the nature of mathematical proof, quadratic equations, pairs of linear algebraic equations, geometric progressions, proof by induction, manipulation of inequalities of real numbers, coordinate geometry, sketching graphs, and (in one chapter) some trigonometry. This may be all that is required, but the elementary nature of this material does not mean that the book is an easy one to work through. Experience in the classroom shows that each of the five main chapters is challenging for the recent high school graduate. Therefore, in order to ease the transition from high school to practical usage, we have included a set of preliminary exercises for each of Chapters 2 through 6. Although they are placed at the end of each chapter, many readers may wish to work through them before embarking on a serious attempt to comprehend the body of the chapter. These preliminary exercises will serve as a refresher for the appropriate high school mathematics and will also pave the way for a ready understanding of the content of the chapter and the exercises thereon.

Five subject areas have been chosen which can be treated in the framework of the mathematical syllabus we have outlined. At the same time, we hope the reader will find these subjects to be interesting and significant as illustrations of the usefulness of mathematics in the real world.

In Chapter 1, we discuss the unifying theme behind the five subsequent chapters, namely, the idea of a mathematical model. The limitations of our knowledge and abilities are generally such that we can rarely predict or explain natural phenomena exactly. We therefore resort to models of the situation and investigate *their* behavior mathematically. The transference of notions from the real world to the model and back again is a vital part of applied mathematics which poses its own problems. We have tried to emphasize these in the text but, due to the difficulties inherent in the development of problems which are both realistic and of mathematical interest, these problems do not play a great role in the exercises. These are, with few exceptions, routine exercises designed to fix ideas and test the reader's understanding of the text. As in most mathematics books, the reader cannot expect to obtain a complete understanding of the contents without making a serious attempt to complete most of the exercises.

The reader will observe that Chapters 2 and 4 are technique-oriented. That is to say, the unifying feature of each chapter is a mathematical technique which may find application in a wide variety of real-world problems. In contrast, Chapters 3, 5, and 6 each have the study of a rather special physical phenomenon as their objective. Progress is made in these chapters by making successively deeper studies of the phenomenon in question and developing the necessary mathematical techniques as we proceed.

It is not claimed that the treatment of the five problem areas is in any way complete. On the contrary, in each case, the problem area is merely opened up and questions of more sophisticated models or analysis are posed without solution. This is part of a deliberate attempt to leave the reader with his interest aroused and with some stimulus to go on to further study in applied mathematics. Even with these limited objectives, however, we are able to bring out the role of a few pervasive ideas which are found in a great variety of problems of applied mathematics. For example, after working through this book, the reader should have some feeling for the ideas of stability, error analysis, linearization, and optimization.

With one exception, Chapters 2 through 6 are independent of one another and can be studied in any order. The exception is Section 4.6, in which it is convenient to use some of the results on recurrence relations obtained in Section 3.10.

The second half of Chapter 6 is perhaps the most difficult, but also the most exciting, part of the book. If the reader does not feel up to this challenge, Sections 6.7 to 6.11 can be omitted without prejudice to the rest of the book. No apology is made for including this introduction to relativity. Experience suggests that many students enjoy this challenge and it does constitute the prime example of the process of re-examination and refinement of mathematical models. It also provides an opportunity to introduce some historical perspective into our discussion of applied mathematics.

It is a pleasure to record my gratitude to several colleagues who have assisted me with constructive criticism and with ideas for exercises. These include Dr. P. A. Binding, Dr. D. S. Watkins, and Dr. D. R. Westbrook all of whom became involved by teaching from one of the several lecture-note versions of this text. I am also indebted to several of the secretarial staff of the Department of Mathematics, Statistics and Computing Science, University of Calgary, who contributed to the production of these versions and to the final typescript.

PETER LANCASTER

CHAPTER ONE

MATHEMATICAL MODELS

1.1 Introduction

Why is it important that students study mathematics throughout their school careers? There are at least two good answers to this question, but let us focus our attention on one answer, namely, that mathematics is useful. If the usefulness did not extend beyond mere arithmetic—the calculations needed to check interest rates or the change from a twenty dollar bill—we would not need to worry about training in mathematics beyond the age of fourteen or fifteen. The fact is, however, that we live in a highly complex society, and it is becoming ever more important for the educated person to have some knowledge of what mathematical science can and cannot do for us. Otherwise, we may well be subject to manipulation by those who seem to be wizards of space research, economic planning, statistics, computing science, and the like.

The purpose of this book is to give the reader some idea, at the level of high school mathematics curricula, of just what applied mathematics is—at least in the author's view. It must be understood that one cannot do applied mathematics without mathematics. Confidence and facility with mathematical techniques are essential, and these, of course, demand some grasp of fundamental mathematical concepts. The techniques required vary from one situation to another, and new applications often demand that new techniques be developed. Although the techniques we shall require are confined to elementary mathematics, it is the author's aim to present applied mathematics as an interesting and stimulating subject which is worthy of your further study.

In this book, we meet with several different areas of application, each of which requires different techniques. There is, however, an important unifying feature in them and in a great many other applications of mathematics. This is the idea of a *mathematical model* for a real-world situation. We will develop this idea in this preliminary chapter.

1.2 The Process of Idealization

In a typical problem of applied mathematics, there is a certain phenomenon of the real world to be investigated. Certain characteristics of this phenomenon are of particular interest; they must be explained and, if possible, their future behavior predicted. A variety of problems are amenable to mathematical analysis; the phenomenon in question may be vibrations in the structure of an aircraft, the propagation of information through biological cells, the distribution of gasoline from storage to retail stations, or the prediction of trends in the national birthrate.

As we concentrate on the problem at hand, we soon find that a new point of view is required. Real-life phenomena are generally so complicated in relation to the mathematical methods at our disposal that we cannot hope to represent and account for their *every* characteristic. Consequently, some simplifying hypotheses must be made. The moment that we do this, we are leaving the real world and beginning to make a (mathematical) model.

This point of view that we must adopt goes right to the foundations of mathematics itself and to geometry, in particular. It is not necessary to be a mathematician to have some appreciation of the ideas of point, straight line, and circle, for example. These ideas are intuitively acceptable to everyone as concepts which can be realized quite effectively by making the appropriate marks on a piece of paper with pencil, ruler, and a pair of compasses. This may be described as the *physical* way of thinking.

For the mathematician, the physical way of thinking is merely the starting point in a process of abstraction or idealization. Instead of being a dot on a piece of paper or a particle of dust suspended in space, a point becomes, in the mathematician's ideal way of thinking, a set of numbers or coordinates. In applied mathematics we must go much further with this process because the physical problems under consideration are more complex. We first view a phenomenon in the physical way, of course, but we must then go through a process of idealization to arrive at a more abstract representation of the phenomenon which will be amenable to mathematical analysis.

Having agreed that idealization is necessary, we must give some thought to how far we can go with the process. Clearly, it is impossible to be specific about this unless we have a particular problem to discuss. However, we hope that, in any problem we tackle, we can simplify and idealize to such an extent that (a) we retain all the relevant features of the phenomenon and strip away all irrelevant or negligible features, and (b) we arrive at a problem which, when certain laws expressed mathematically are imposed, has a solution we are able to find.

We shall discuss these two aspects of the process separately. If, in considering (a), we try to retain too many features of the problem, we may arrive at an intractable or insoluble mathematical problem. If we retain too few, we lose all contact with reality and our mathematical results will be meaningless. It can be seen that the process of arriving at an appropriate model may often be difficult. It may depend on a process of trial and error or on accumulated experience with similar problems. It also depends to a large extent on physical intuition and on common sense.

"To illustrate, suppose we are interested in the periodic time of a bar suspended from one end, and oscillating as a pendulum. Which properties of the bar is it essential for us to bear in mind, and which may we neglect as incidental? Can we predict the periodic time of oscillation without knowing the material of which the bar is constructed? Does the form of the cross section of the bar matter? Does it make any difference whether the bar is supported on a knife-edge or by bearings? The cautious, well-informed physicist would say that all these things mattered and many others. One material yields more than another, the form of cross section influences the distribution of material, and a change in the mode of suspension may alter the axis about which the pendulum oscillates. But if we were as cautious as this, we should have no science of mechanics. To start on the problem at any rate, we must simplify it ruthlessly. So we think of the bar as a rigid mathematical straight line, and the support, as a fixed mathematical point. Now we have a problem which is reasonably simple to handle mathematically. Strictly speaking, no properties are incidental. Even the color of the bar affects the pressure of light on it; a subway train stopping five hundred miles away may cause a vibration in the support and affect the motion of the bar. Common sense, which is the accumulated experience of centuries, gives us some guide as to the factors which we may neglect."*

Even though the problems of idealization faced above may seem to be severe, they are simple when compared to many of the situations faced in biology or economics. In the case of the oscillating bar, our intuition gives a reliable guide as to which of the influences mentioned will be significant. If we attempt to set up a model of interacting animal species (possibly competing with one another, or one being a prey for another) it becomes rather more difficult to make sensible decisions concerning those factors which can and cannot be safely ignored. How are the populations to be measured? How are the rates of reproduction affected by food supply? Should fluctua-

*Quotation from Synge and Griffith (1959).

tions in climate and the social structure of a species be accounted for? In such cases the precision of conclusions drawn from the model is likely to be in question. In spite of this, valuable qualitative information concerning the interaction of different effects and their relative importance may be obtained.

It is clear that, in this process of idealization, a thorough understanding of the physical nature of the problem at hand is required. For this reason, it may often be necessary for the applied mathematician to consult with an engineer, economist, biologist, as the case may be, in arriving at a model which is acceptable to the specialists of the science in question.

1.3 The Model Completed

In the previous section we focused attention on the first part of the model-making process, that of idealization. This idealization is achieved by making simplifying assumptions concerning the nature of the phenomenon in question, and by assuming that certain physical influences on the problem will have such a small effect that they can be neglected. The next step of the process consists in imposing certain mathematically defined laws on the idealized problem. There are generally certain unknown quantities which we are to calculate or to estimate. The laws we impose involve relationships between these unknown quantities which may take the form of equations or inequalities. Where do these laws come from? Again, we cannot be precise about this without being more specific. However, the laws are generally derived from principles governing the scientific field to which we are applying our mathematics. For example, in the oscillating-bar problem discussed above, the laws imposed could be Newton's laws of motion.

These are postulates which were formulated after many years of study and observation of the phenomena of mechanics and have been very convincingly confirmed by experiment. In the more recently developed fields of application such as the social and biological sciences, the laws are not so highly structured, perhaps because they have not been discovered yet. In these areas, the laws and conditions to be imposed may have to be obtained from a direct intuitive assessment of the problem.

At this stage we must stand back and take an overview of our model-making. Having idealized and imposed laws and conditions, we would hope that we have now arrived at a mathematically defined problem that we can solve. In order to do this we should have some confidence that (a) the laws and conditions imposed are consistent with one another, i.e., they should not be mutually contradictory, and (b) they are sufficient in number and scope to determine a solution. Our ability to ensure these conditions will depend on our knowledge of the field of mathematics appropriate to the formalism of the model. In the jargon of mathematics, conditions (a) and (b) can be interpreted as demands for existence and uniqueness theorems.

If we have some confidence that (a) and (b) are satisfied, then we can say that the mathematical model is complete. What do we do now? Since we are presuming that a well-posed problem about mathematics has been set up, we now go ahead and solve it. In this step of the process we are essentially doing mathematics. We apply the logical arguments of analysis, algebra, etc., to find a solution to the problem. In many cases the real-world problem which gave rise to the model can be left aside at this stage, and we can operate in the comfortably logical, albeit demanding, world of pure mathematics. But even at this stage it is surprising how often it helps to keep in mind the physical significance of the logical steps that may be taken. For example, knowledge of the physical implications of an intermediate stage of the analysis may give a clue to the next line of argument.

At this point it can also be remarked that the special problems posed by applications often require special techniques and methods for their solution. We enter a gray area here between applied and pure mathematics, as the author understands them. It is essentially mathematics that is to be done, but the purist is not interested because he does not see the structure and abstraction on which he thrives. This kind of developmental mathematics is often the lot of the applied mathematician. He is happy to do it because he can see a use for it, and this is the mainspring of *his* motivation.

Since the applied mathematician must very often resort to numerical solution of the model problem, he should also be familiar with numerical analysis and computers. Thus, a second gray area develops between applied mathematics and computing science. Since the study of widely useful methods of mathematics and of numerical analysis are of interest primarily to the applied mathematician, a case is made for their inclusion under the general heading of applied mathematics.

Having laid claim to the mathematical and numerical methods likely to yield solutions, let us now suppose that they have been applied to good effect and a solution is available to the mathematically posed model problem. What does this do for us? Well, it tells us something about the model—this abstraction which exists only in the mind of the applied mathematician. There is a large step to be taken in interpreting the meaning of the results back in the real-world situation once more. This is generally an easy step, since we should be clear about those things in the real-world problem which correspond to the characteristics of the model problem. But note that, at best, the result can be true only for the model and can only *suggest* what may be true for the real-world problem. We shall return to this point in the next section.

If the result of our analysis is a mere suggestion, how can we establish any confidence in the technique of model-making and analysis? The answer constitutes the final step of the process and its only justification. We must now compare the results suggested for the real-world problem with experiment. This comparison may be informal in the sense that the results obtained are obviously consistent with earlier experience, or it may be very formal in that experiments must be carefully designed and performed under laboratory conditions. At the same time, the human mind has great tenacity, and one's belief in particular models or theories can be inordinately strong in the face of conflicting (or even contradictory) evidence. This kind of tenacity may lead to great discoveries but may also help to explain the long life of the flat-earth and geocentric theories of antiquity.

Let us now summarize the steps in this general procedure of applied mathematics:

1. Attention is focused on a phenomenon of the real world. Certain features or patterns of behavior are to be explained or predicted.
2. A model is constructed in two steps:
   (a) The phenomenon is idealized.
   (b) A sufficient number of consistent conditions are imposed.
3. Mathematical reasoning is applied to obtain appropriate conclusions concerning the model.
4. The conclusions concerning the model are interpreted in the real-world situation.
5. The results are compared with observation.

In the terminology of M. S. Klamkin (1971), step 1 is described as "Recognition", step 2 as "Formulation", step 3 as "Solution and Computation", and steps 4 and 5 as "Explanation".

1.4 Truth and Validity

It is now clear that whatever predictions we make about the real-world based on an analysis of a mathematical model are always limited by the hypotheses of step 2 above. It may be said that we are playing a game of make-believe so that no absolute truth can result from our work. We say that *if* (H) the hypotheses of steps 2(a) and 2(b) are true, *then* (C) the conclusions of step 4 are true. In abbreviated form we have the statement: $H \Rightarrow C$. As in any experimental science, repeated confirmation of our conclusions C at step 5 will increase our confidence in the hypotheses H. But we shall never be able to say that $C \Rightarrow H$. That is, repeated agreement of predictions with experiment will not allow us to say that the hypotheses are *true* in an absolute sense. We can, however, talk about the *validity* of a mathematical model subject to our hypotheses. Validity means that a convincing number of experiments assure us that the conclusions of our analysis conform with the real-world phenomenon to within some desired accuracy.

So much for the fortunate situation in which the predicted conclusions are acceptable when interpreted in the real world. What if they are not acceptable, or if it is found at step 5 that the predicted conclusions do not agree with observation? We now re-examine the model-making steps in the reverse order. Checking step 4 should again give us no trouble. Step 3 will often warrant very careful attention, but we must assume that, with sufficient care, step 3 will be correctly performed.

In practice, it is very often the case that faults appear in step 2. In order to make problems tractable, there is some pressure to impose seemingly natural laws which are just too simple to represent the problem adequately, or to ignore some effect which, if included, will give rise to a brute of a mathematical problem. In such cases the discrepancy between results and observations may well suggest where the error lies. If so, this may be modified to provide a new model, and we start the process again.

In Chapter 6 we discuss the transition from classical (Newtonian) mechanics to the theory of special relativity. In this instance we get a beautiful example in which the most primitive assumptions of a theory (measurement of time and distance) had to be re-examined in order to produce a model which would agree with certain experiments. As we have just remarked, neither model can be said to be *true*, but both are found to be *valid* within the respective sets of hypotheses laid down in step 2.

In spite of our remark concerning the liability to err in step 2 of the model procedure, it is a continual source of surprise that plausible conclusions can often be found on the basis of very crude assumptions. This good fortune and plain common sense both suggest that, when faced with a particular problem, we should first examine a very crude and apparently oversimplified model in order to arrive quickly at a preliminary conclusion. We then refine the hypotheses, perhaps in two or three steps, and solve successively more difficult model problems in each case. The results of one model will then serve as a check on those of its successors.

As a final comment on our apparent good fortune, it may also happen that physical laws and theories are suggested by mathematical analysis! In the pressure of confusing (and even apparently conflicting) physical evidence, a good mathematical formulation of a problem may lead to a reassessment of the natural laws invoked and to the identification of the most important variables involved. The discovery of some of the elementary particles of physics has been strongly influenced by this kind of investigation.

1.5 An Illustrative Example of Inventory Control

To complete this chapter, we shall examine a particular problem in the light of our discussion of modeling. In a manufacturing process, a ton of a certain material is used per day on the factory site and 100 tons are stored some distance away in a warehouse. It costs money to maintain the warehouse storage facility, but it costs considerably more to store material on the factory site where space is at a premium. There are also costs associated with shipment of material from warehouse to factory. The manufacturer may then be interested in the following problem of inventory control: In what quantities should the material be moved from warehouse to factory in order to minimize the total cost, C, of storage and transport for the 100 days of operation? We shall assume that k tons are to be moved at a time at intervals of k days.

In formulating the question, we have already begun with the "Recognition" phase, step 1, of the modeling process. We proceed to step 2. How does the manufacturer estimate the storage costs? This is where the process of idealization will begin to play a significant role. For the purpose of this study, the estimate may include amortization of building costs, salaries, maintenance of buildings, and so on. In general, these items can only be estimated, and in operation there will always be uncertainties due to irregular deliveries, strikes, illness, etc. In spite of all these factors, the accounting staff will idealize their plant and process and arrive at representative figures.

We shall assume that, where warehouse space is not being used for the material we have in mind, it is used to store something else, in which case a storage cost per ton per day is sought. Suppose we are given a figure of $2 per ton per day at the warehouse. In contrast, the cost at the factory is proportional to the *maximum* amount held there which, as indicated above, is k tons. We suppose that we are given $10k$ per day for storage at the factory. (Thus, the minimum costs for storing one ton for one day at the warehouse and at the factory are $2 and $10, respectively.)

Since transport is provided in large units of fixed capacity, transport costs are set per *shipment* rather than per ton: say $144 per shipment, independent of the value assigned to k.

We now assume that the data given provides a reasonable first model. There may be differences in cost due to time lags in delivery and other effects, but we assume that, in the eyes of the manufacturer, these are judged to be relatively insignificant and are disregarded as part of the idealization process. This model is now used to find a first estimate of that value of k for which C, the total of storage and transport costs, is least.

We now begin with step 3 and formulate C, the total cost, in terms of k. For this, we shall need the formula for the sum of the first n integers:

(1) $$1 + 2 + 3 + \cdots + n = \tfrac{1}{2}n(n + 1).$$

For convenience, suppose we first confine our attention to those k for which $100/k$ is an integer, so that there are precisely $100/k$ shipments to be made. If k tons are moved out of the warehouse on the first day, then there will be $(100 - k)$ tons stored there for the first k days at a cost of

$$\$(100 - k) \times k \times 2 = \$2k^2\left(\frac{100}{k} - 1\right).$$

For the next k days the number of tons to be stored is reduced by k so that, for this period, we get a reduction in cost of $\$2k^2$, and the cost for this period of k days is

$$\$2k^2\left(\frac{100}{k} - 2\right).$$

Proceeding in this way and summing up the costs for each of the $100/k$ periods of k days, we get the total cost of storage at the warehouse in the form:

$$2k^2\left(\frac{100}{k} - 1\right) + 2k^2\left(\frac{100}{k} - 2\right) + \cdots + 2k^2(2) + 2k^2(1)$$

$$= 2k^2\left[1 + 2 + 3 + \cdots + \left(\frac{100}{k} - 1\right)\right]$$

$$= 2k^2\left[\frac{1}{2}\left(\frac{100}{k} - 1\right)\frac{100}{k}\right]$$

using Eq (1). This expression simplifies to $\$(100 - k)100$ for the total warehouse costs.

The costs at the factory site for the 100 days are simply $\$(10k)100$ and, finally, the transport costs are $\$144 \cdot (100/k)$. Thus, the total costs (in dollars) of storage and shipping are

$$(2) \quad C(k) = 100\left[(100 - k) + 10k + \frac{144}{k}\right] = 100\left[100 + 9k + \frac{144}{k}\right].$$

This function, C, of k contains the information needed to solve the model problem. If we examine the graph of $C(k)$ in Fig. 1.1, it appears that C is least when $k = 4$. This is indeed the case and we will not concern ourselves with the mathematical niceties of proof since we wish to focus only on the modeling problem.

This model suggests, therefore, that the material be shipped in loads of 4 tons at 4-day intervals and, happily, 4 divides 100 so that there are exactly 25 shipments. The total cost C is predicted to be $\$172,000$, compared to $\$253,000$ if the material is shipped 1 ton per day, and $\$190,000$ if shipped in 8-ton loads at 8-day intervals. One must then ask whether the solution is realistic (step 4). Is this conclusion consistent with the standard capacity of the units of transport? Can four tons be accommodated at the factory? In

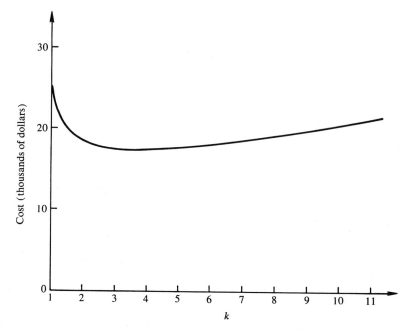

Figure 1.1 Total cost as a function of k.

the light of this conclusion, should the hypotheses made in setting up the model be re-examined? These and many other questions may have to be answered before our solution of this model problem could be deemed to provide a practical solution to the real-world problem. As discussed in Section 1.4, it is clear that the prediction of the model could not be described as "true" but, depending on the answers to the kind of questions posed above, the model may well have validity.

Exercises

This set of exercises is probably best used as a basis for classroom discussion.
1. How would you idealize the earth when considering:
   (a) the motion of the earth relative to the sun?
   (b) the course of a ship across the Pacific?
   (c) the speed of a helicopter relative to the earth?
   (d) the orbit of an earth satellite?
2. Consider the example of Section 1.5 once more.
   (a) If the transportation costs are quoted as $140 per shipment instead of $144, how would the conclusions be modified?
   (b) If the transportation cost is $81 per shipment, verify that the minimum point on the new curve for $C(k)$ is at $k = 3$. Then $100/k$ is not an integer. What

would you recommend to the manufacturer in this case? How must the analysis for step 3 in Section 1.5 be modified to get a precise solution for the model problem? Is it worth the effort?

3. Imagine that you are hired by the country of Bleatar to forecast the changes in distribution of population in the country as a whole and in its two-only urban centers. The study is to cover the next ten-year period. What sort of data would you seek from the Department of Immigration and the Bureau of Statistics? Discuss the formulation of a mathematical model for the problem.

4. How would you idealize an egg when asked to model the cooking process as the egg is boiled?

CHAPTER TWO

LINEAR PROGRAMMING

2.1 Introductory Example

In this chapter we shall develop some primitive examples of a modern technique which is widely used in applied mathematics. The limited mathematical techniques at our disposal will confine our detailed analysis to some problems which are very special and, admittedly, rather artificial. However, the reader may get a feeling for the type of problem which is amenable to solution by the method of linear programming and, at the end of the chapter, we indicate how more general problems may be tackled.

Before starting any mathematical analysis, let us look at a particular example. This concerns a cotton-goods manufacturer who has 1000 yards of cotton at his disposal to make shirts and pajamas. He needs 1 yard to make a shirt and 2 yards for a pair of pajamas. Thus, if he makes x shirts and y pajamas, the limited amount of material imposes a *constraint* (or condition) on his production as follows:

$$(1) \qquad\qquad x + 2y \leq 1000.$$

We shall impose additional constraints. The manufacturing process can be broken down into two parts: "cutting" and "making up." The labor available and the efficiency of the two processes are responsible for the further constraints. The cutting process requires 2 hours per shirt and 1 hour per pajama suit, so that the total time used in this process is $2x + y$ hours. However, there are only 1200 hours of labor available for cutting, and therefore x and y must be chosen so that

(2) $$2x + y \leq 1200.$$

If making up requires 2 hours per shirt and 3 hours per pajama suit, and there are 1600 hours available for this process, then we have

(3) $$2x + 3y \leq 1600.$$

Obviously, we also have

(4) $$x \geq 0 \quad \text{and} \quad y \geq 0.$$

The conditions (1)–(4) make up the constraints on the problem. However the manufacturer decides on the number of shirts, x, and the number of pajamas, y, this pair of numbers must simultaneously satisfy the five constraints.

The manufacturer needs some criterion by which to determine x and y. We claim that he can do this in such a way that his *total profit* will be maximized and, simultaneously, the five constraint conditions will be satisfied. Obviously, his total profit depends on x and y. We shall assume a functional dependence and write $P(x, y)$ for the total profit if x and y pajamas are made.

Suppose he knows that the profit per shirt is exactly the same as that for a suit of pajamas. His objective then is to choose x and y so that the conditions (1)–(4) are all satisfied *and* the quantity

$$P(x, y) = x + y$$

is as big as possible.

The number pairs (x, y), if any, for which all the constraints are satisfied:

(5)
$$\begin{aligned} x &\geq 0, \quad y \geq 0 \\ x + 2y &\leq 1000 \\ 2x + y &\leq 1200 \\ 2x + 3y &\leq 1600 \end{aligned}$$

are called the *feasible solutions* of the problems. The function $P(x, y)$ is called the *objective function*, and any feasible solution (x, y) which yields the maximum possible value for the objective function is a solution of the problem.

Notice that the constraints and the objective function involve only *linear* functions of x and y (see Exercises 1 and 2). For this reason, this is called a *linear* programming problem. If either the constraints or the objective function involve functions that are not linear, then we have a nonlinear programming problem.

There is a rather special property of this example. Notice that, by our definition, x and y should be integers. However, this is not an important feature of the problem. If we simply go ahead and treat x and y as real numbers and if a solution is not a pair of integers, we merely replace x by the largest integer less than x, and similarly for y. There *are* problems where only integer solutions are acceptable and more care needs to be taken to ensure that we get the best solution *in integers*. Such a problem is called an *integer programming* problem.

Now let us consider this problem from the point of view of "model-making". We have built a model for what might well be a real-life situation and, in doing so, we have used certain idealizations and imposed certain laws. In our problem, only the (hypothetical) manufacturer could discuss these critically. We are supposing that the manufacturer does, indeed, find the model a plausible one. For example, we assume that he has confidence in the hypothesis that, say, the profit is the same on one shirt as on a pajama suit. The manufacturer may be aware that this can only be an approximation, but his calculations and experience lead him to believe that this is of no concern. We shall see that, in this problem, this particular question is *not* significant, and precisely the same solution would be obtained if the profits differed by as much as 30%. Note also that we have assumed that the profit on shirts, say, is proportional to the number made. We shall argue in Section 4.1 for quite a different relationship between profit and production.

Equally well, the constraint $2x + y \leq 1200$ imposed by the labor and time available for the cutting process can be only an approximation to the real situation. The figure of 1200 hours available for the process may well be a very rough estimate subject to the availability of labor and machinery for the production line. Clearly, if the model implies conclusions which are not acceptable to the manufacturer for one reason or another, then all of these hypotheses would have to be carefully scrutinized and possibly modified to formulate a new model.

Exercises for Section 2.1

1. Let a, b, c be fixed real numbers. We can defined a real-valued function f of three variables x, y, z by assigning

$$f(x, y, z) = ax + by + cz.$$

Prove that, for any real number k and any real x, y, z,

$$f(kx, ky, kz) = kf(x, y, z).$$

Prove also that, for any real numbers $x_1, y_1, z_1, x_2, y_2, z_2$,

$$f(x_1, y_1, z_1) + f(x_2, y_2, z_2) = f(x_1 + x_2, y_1 + y_2, z_1 + z_2).$$

These properties show f to be a *linear* function.

2. Let f be a function of several variables, say $x_1, x_2, \ldots, x_n$. Suppose that f is defined for all real values of the x's and that the values taken by f are real numbers. The function f is defined to be a linear function if, for every real $k, x_1, x_2, \ldots, x_n$, we have

$$f(kx_1, kx_2, \ldots, kx_n) = kf(x_1, x_2, \ldots, x_n)$$

and for every real $x_1, x_2, \ldots, x_n, y_1, y_2, \ldots, y_n$ we have

$$f(x_1, \ldots, x_n) + f(y_1, \ldots, y_n) = f(x_1 + y_1, \ldots, x_n + y_n).$$

(a) Using the preceding exercise as a guide, write down some linear functions of the kind described.
(b) Prove that if a, b, c are real, $c \neq 0$, and f is defined by

$$f(x, y, z) = ax + by + c$$

then f is *not* a linear function.
(c) If f is defined by $f(x) = x^2$, prove that f is not a linear function.

2.2 Linear Inequalities

In analyzing problems like the one above, we must be very familiar with the solutions of linear inequalities of the form

(6) $$ax + by \leq c,$$

where a, b, and c are given real numbers and not both of a, b are zero. The *solution set* of such an inequality is the set of all ordered pairs of real numbers (x, y) for which (6) is a true statement. In set notation:

$$\{(x, y): \quad ax + by \leq c\}.$$

Each such number pair corresponds to a point in a plane in which rectangular coordinate axes are defined. Using this correspondence, we describe the solution set geometrically. We know that the solution set for the *equation* $ax + by = c$ is a straight line in the coordinate plane. We *claim* that the solution set of the condition $ax + by < c$ is the set of all points in the plane on *one side* of this line but not including points on the line. The other side of the line is the solution set for $ax + by > c$. Each of these sets is known as an *open half-plane*. The solution set for $ax + by \leq c$ is then the union of an open half-plane and the points on the defining line itself. This is called a *closed half-plane*.

Let us indicate a formal proof of these facts. Consider first any condition $ax + by \leq c$ with $b > 0$. Then this condition and

(7) $$y \leq mx + d$$

are equivalent if $m = -a/b$, $d = c/b$. Then $y = mx + d$ is the equation of a line in the familiar "slope-intercept" form: m is the slope and d the intercept on the y axis.

Take any point (x_0, y_0) not on the line, and draw a line through this point with slope m. The equation of *this* line must now be of the form $y = mx + d_0$, where d_0 is *its* intercept (see Fig. 2.1). We may characterize all points on one side of the line as those which give rise to an intercept $d_0 > d$. For all points (x_0, y_0) on the *other* side, we obtain $d_0 < d$.

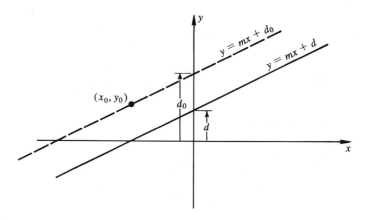

Figure 2.1 Lines with slope m.

Let (x_0, y_0) be any point in the plane on that side of the line for which $d_0 > d$. Then, since this point is on the line with equation $y = mx + d_0$, we have

$$y_0 = mx_0 + d_0 > mx_0 + d.$$

Thus, all points on this side of the line are in the solution set of the condition $y > mx + d$, that is, of the condition $ax + by > c$. Conversely, we can argue that $ax_1 + by_1 > c$ implies $d_1 > d$, so that (x_1, y_1) is on the same side of the line as (x_0, y_0). Similarly, we show that a point (x_0, y_0) is on the other side of the line $y = mx + d$ (where $d_0 < d$) if and only if we have $ax_0 + by_0 < c$.

We have now substantiated our claim in the case $b > 0$. If $b < 0$ then (6) is equivalent to $y \geq mx + d$, and a similar argument can be employed to prove the result in this case. We leave the case $b = 0$ as a trivial exercise for the reader.

In practice, we need a technique for deciding which side of the line $ax + by = c$ corresponds to the inequality $ax + by \geq c$ and which side to $ax + by \leq c$. We can do this by choosing a particular test point (x_0, y_0) which is known to lie on one side, calculating $ax_0 + by_0$, and comparing this with c. If $ax_0 + by_0 \geq c$, then we know that the side on which (x_0, y_0) is known to lie corresponds to the solution set of $ax + by \leq c$, and therefore the other side corresponds to $ax + by \geq c$.

For example, consider the condition $x + 2y \leq 4$. In Fig. 2.2(a) we have sketched the line $x + 2y = 4$. Consider the origin as test point $x_0 = 0$, $y_0 = 0$. For this choice we have $ax_0 + by_0 = 0$ which is less than 4. Thus

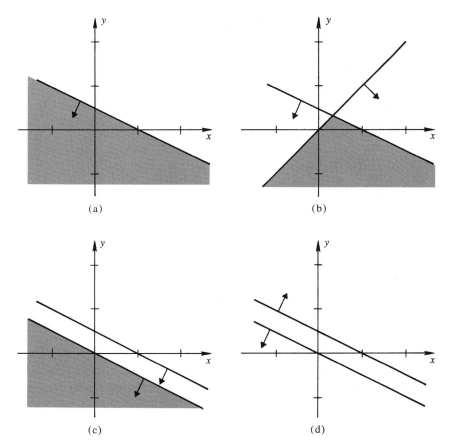

(a)

(b)

(c)

(d)

Figure 2.2 Solution sets in the plane.

the closed half-plane in which the origin lies (hatched region) corresponds to the solution set of $x + 2y \leq 4$. We may also indicate the appropriate half-plane merely by attaching an arrow to the straight line pointing to the side on which the condition is satisfied.

In linear programming problems, we generally have several inequalities of the form in (6) to satisfy simultaneously. Thus, the solution set (the set of feasible solutions of a linear programming problem) may be described geometrically as those points (if any) common to several closed half-planes, one half-plane corresponding to each inequality.

In the case of two inequalities, where the two linear equations in question represent lines which are not parallel, the solution is an infinite sector of the plane. In Fig. 2.2(b) the hatched region corresponds to:

$$\{(x, y): \quad x + 2y \leq 4 \quad \text{and} \quad x - y \geq 0\}.$$

If the two lines are parallel but not coincident, the solution set may be a closed half-plane or may contain no points at all. In Fig. 2.2(c) we illustrate the case of $x + 2y \leq 4$ and $x + 2y \leq 0$. In this case the first condition is redundant for, if the second condition is satisfied, the first is automatically satisfied. In Fig. 2.2(d) we have $x + 2y \geq 4$ together with $x + 2y \leq 0$, and the solution set is empty; the conditions are contradictory.

Exercises for Section 2.2

1. Draw sketches to indicate solution sets of:
  (a) $x \geq 1$ (b) $-1 \leq x \leq 4$
  (c) $5x + 7y \geq 70$ (d) $x - 3y \leq 1$
  $ \qquad\qquad\qquad\qquad\qquad\qquad x \qquad\;\; \geq 0$

  (e) $3x + 2y \leq 6$
  $ \quad x - \;\; y \geq 1$
  $ \quad x \qquad\;\; \geq -1$
2. Find a set of constraints whose solution set is the interior (including edges and vertices) of the triangle with vertices $(0, 0)$, $(1, 2)$, $(-2, 4)$.
3. Find the set of constraints whose solution set is the semicircle (including perimeter) with center $(0, 0)$ and radius 2 and bounded on the right by the y axis.

2.3 The Objective Function

We have seen above that the constraints of a linear programming problem will generally determine a set of feasible solutions which can be represented by a region of the x-y plane bounded by line segments, lines, or half-lines. In a programming problem (linear or otherwise), we have a function of x and y, say $f(x, y)$ whose value is to be made either as large or as small as possible as x and y vary in the set of feasible solutions. As in the example in Section 2.1, f is called the *objective function*.

We suppose that the objective function always has the form

(8) $$f(x, y) = c_1 x + c_2 y$$

for some real c_1 and c_2 (see Exercises 1 and 2 of Section 2.1), and the objective is to find a feasible solution (x, y) which makes $f(x, y)$ as large as possible. In some problems we are asked to make $f(x, y)$ as *small* as possible, but we can include that case in the present discussion merely by noting that the problem of finding the minimum of $f(x, y)$ is essentially the same as that of finding the maximum of $-f(x, y)$.

A graphical procedure can be used for the maximization problems we shall meet in the next section. Suppose we first assign an arbitrary constant value for f, say k_0. Then

$$c_1 x + c_2 y = k_0$$

is the equation of a straight line. We draw the graph of this equation and observe whether it intersects with the region in which the feasible solutions lie. If not, we discard this value of k_0, for we are only interested in values of the objective function obtained from feasible solutions. If we choose a new value of f, say k_1, we obtain a line *parallel to* that for k_0 but shifted to right or left, depending on the difference $k_1 - k_0$. Indeed, by assigning all possible real values to k in the equation $c_1 x + c_2 y = k$, we obtain a *family* of parallel straight lines which fills the whole plane.

We are interested only in those values of k which yield straight lines having at least one point in the region of feasible solutions. If we are to *maximize f*, then we seek the *largest k*, say k_c, such that the line $c_1 x + c_2 y = k_c$ contains a feasible solution (see Fig. 2.3). Each time we assign a numerical

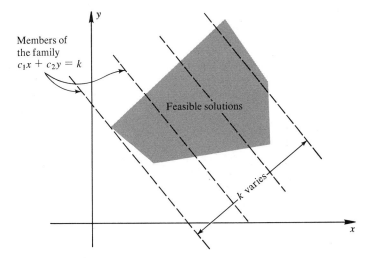

Figure 2.3 The optimization procedure.

value k to the function f and sketch the graph of $f(x, y) = k$, we determine a *contour* or *level curve* of f. Thus, our procedure is to examine the nature (and particularly the levels) of the contours of f within the region of feasible solutions. The procedure will be clarified by two worked examples.

It is geometrically clear that, in certain cases, we can apply the same kind of procedure even when the boundaries are curves and not straight lines, or even when the objective function is nonlinear. Some examples of this kind are included in the exercises at the end of Section 2.4.

2.4 Two Examples

Consider first the example of Section 2.1. The constraints are given by the inequalities (5). The region of feasible solutions is hatched in Fig. 2.4. We now examine the objective function, $P(x, y) = x + y$. If we presuppose a profit of 500 units, then $x + y = 500$, and we observe that the graph of this line does indeed cross the region of feasible solutions. Thus, any (integer) feasible solution (x, y) on this line represents a number of shirts, x, and pajamas, y, which if manufactured will yield a profit of 500 units. But we can do better than this. If $P(x, y) = 600$, we obtain a line which still has feasible solutions. We now go on increasing k in the equation $x + y = k$ beyond 600. As we do so, we generate more members of the family of parallel lines (moving to the right in Fig. 2.4) and can go on doing so until we reach the line which passes through the vertex A. If we attempt to increase k (the profit) any further, we obtain a line that does not touch the region of

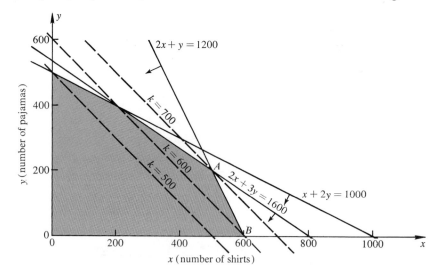

Figure 2.4 Shirts and pajamas.

feasible solutions. That is, there will be no feasible solutions for which that profit can be attained.

It is now clear geometrically that the maximum of f is obtained at point A. We determine the coordinates of A by solving the simultaneous equations obtained from the lines passing through A. If $A = (x_c, y_c)$, we have

$$2x_c + 3y_c = 1600$$
$$2x_c + y_c = 1200.$$

These equations have the solution $x_c = 500$, $y_c = 200$. The manufacturer should therefore plan to make 500 shirts and 200 pajamas for a total profit of 700 units.

Notice that, from the formulation of inequalities (2) and (3), the total available time for the cutting and making up processes is absorbed. However, since $x_c + 2y_c = 900$, we see from the first constraint that the manufacturer will have a *surplus* of 100 yards of cotton. If he has already bought the cotton, then this may be lost. On the other hand, if this calculation was done before the order was placed, the manufacturer can reduce the quantity!

Suppose that we change the problem and assume that a shirt brings twice the profit obtained on pajamas. Then the total profit is proportional to

$$P(x, y) = 2x + y.$$

If we consider the family of lines $2x + y = k$, we obtain lines parallel to the edge AB of the region of feasible solutions in Fig. 2.4. It is clear that the maximum value of k is now obtained for *any* pair (x, y) corresponding to a point on the segment AB. Thus if he produces 500 shirts and 200 pajamas (corresponding to point A), or if he produces 600 shirts and no pajamas (point B), he makes the same maximum total profit of 1200 units. We note that the maximum profit is uniquely defined, although the values of (x, y) yielding this maximum are not unique.

In such a situation, the manufacturer would need some criterion for deciding which (x, y) to choose, but this must be a criterion which is not built into our model. For example, he may prefer solution A because this implies that all of the available labor time for the making up process is used. The solution corresponding to B implies a total time of 1200 hours for the making up process and a surplus of 400 hours. He may prefer to have no such surplus in order to maintain good labor relations, but this is not represented in our model.

As a second example, consider the plight of a farmer who must choose between buying two fertilizers. He knows the composition of these products, and they are both made up of a mixture of two basic chemicals with small amounts of other additives which the farmer considers to be unimportant.

On the basis of the area of land involved and the composition of the soil, he calculates that he needs at least 300 tons of the first chemical, say chemical A, and at least 400 tons of chemical B.

The proportions of chemicals A and B in the two fertilizers are indicated in the following table. Fertilizer 1 costs \$10 per ton and fertilizer 2 costs \$7 per ton. How much of each fertilizer should he buy in order to *minimize* the total cost?

	Chemical A	Chemical B	Other Chemicals
Fertilizer 1	55%	40%	5%
Fertilizer 2	20%	70%	10%

Let x, y be the number of tons of fertilizers 1 and 2, respectively, which he buys. Since he needs at least 300 tons of chemical A, we see from the first column of the table that x, y must be chosen so that

$$(0.55)x + (0.2)y \geq 300.$$

Similarly, the constraint on the amount of chemical B yields

$$(0.4)x + (0.7)y \geq 400.$$

Obviously, we also demand $x \geq 0$ and $y \geq 0$. The region of feasible solutions determined by these four constraints is indicated in Fig. 2.5.

The total cost in dollars of x tons of fertilizer 1 and y tons of fertilizer 2 is

$$C(x, y) = 10x + 7y.$$

This is the objective function which is to be made as small as possible by choice of x and y from the set of feasible solutions. We consider the family of parallel lines:

$$10x + 7y = C.$$

The broken lines in Fig. 2.5 indicate the lines of this family for $C = 10,000$, 8000, and 6000. Therefore, to make the total cost C as small as possible, we must choose that value of C corresponding to a line as far left as possible, and this line is that member of the family passing through vertex A of the region of feasible solutions.

Let A have coordinates (x_c, y_c). These coordinates are determined by the solution of the equations

$$(0.55)x_c + (0.2)y_c = 300$$
$$(0.4)x_c + (0.7)y_c = 400.$$

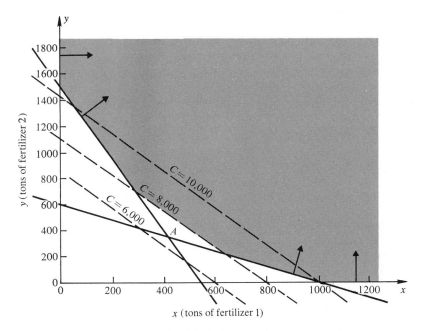

Figure 2.5 The fertilizer problem.

It is found that, to the nearest ton, $x_c = 426$ tons and $y_c = 328$ tons. The total cost in dollars is then

$$(10)(426) + (7)(328) = 6556.$$

This compares with a total cost of $10,000 if he attempts to meet his minimum requirements using only fertilizer 1, and $10,500 using only fertilizer 2.

Exercises for Section 2.4

1. Make a sketch to indicate the square whose vertices are at the points with coordinates (2, 0), (4, 2), (2, 4), and (0, 2). Let S denote the set of points in and on the square.
   (a) At what point of S does the function $P(x, y) = 4x + y$ attain its *greatest* value? What is the maximum value?
   (b) Where does the function $Q(x, y) = x + y$ attain its *least* value in S? What is this least value?
2. Sketch five contours (level curves) of the following functions:

$$x, \qquad 2x - 4y, \qquad x^2 + y^2, \qquad \frac{x}{y}, \qquad \frac{x}{y^2}.$$

3. (a) Find the set, S, of feasible solutions for the constraints

$$x \leq 4, \qquad y \geq 0, \qquad x + 2y \leq 6 \quad \text{and} \quad x - y \geq 0.$$

(b) Find the maximum and minimum values of the function $f(x, y) = 6x + 2y$ attained in S.

4. A farmer has 35 acres of land available on which to grow some sugar beets and some corn. The cost of cultivation per acre, the workdays needed per acre, and the profit per acre are indicated in the table.

	Beet	Corn	Total available
Cultivation cost per acre	$50	$30	$1300
Days of work per acre	2 days	4 days	80 days
Profit per acre	$100	$100	

The acreage he cultivates is limited by the amount of money available for cultivation costs and by the number of working days that can be put into this part of his business, as indicated in the last column.

Sketching graphs as an aid to your calculation, find the number of acres of the two crops which he should plant in order to maximize his profit.

5. Two raw materials, R_1 and R_2, undergo a separation process to produce three minerals M_1, M_2, and M_3. One ton of R_1 produces 0.03 ton of M_1, 0.125 ton of M_2, and 0.4 ton of M_3. One ton of R_2 produces 0.6 ton of M_1, 0.25 ton of M_2, and 0.05 ton of M_3. Sufficient quantities of R_1 and R_2 must be processed to yield totals of *at least* 30 tons of M_1, 25 tons of M_2, and 20 tons of M_3 for the next stage of the industrial process.

If R_1 costs $250 per ton and R_2 costs $200 per ton, how many tons of each should be processed in order to produce the required quantities of minerals and minimize the total cost of raw materials?

6. A confectioner sells two assortments of cookies—"Scrumptious" and "Delicious". Boxes of the former ($2.50 each) are guaranteed to contain at least 20 fig rolls, 40 chocolate chips, and 15 gingersnaps; equivalent figures for "Delicious" ($2.00 a box) are 50 fig rolls, 18 chocolate chips, and at least 10 ginger cookies.

Kids club together and send little Jimmy (who can carry only 10 boxes) to the store. Between them, 20 kids like fig rolls best, 28 like chocolate chips, and 10 prefer gingersnaps. How many boxes of each should be bought so that each child can have at least four of his favorite choice with minimum expenditure?

7. (a) Sketch the solution set D of

$$y \leq x + 1$$
$$y \geq x^2 - 2x + 1$$

(b) What is the maximum on D of (i) $x + y$ and (ii) $-y$.

8. Let D be the solution set of the following conditions:

$$x \geq 0, \qquad y + 2x \leq 10, \quad \text{and} \quad x^2 + y^2 \leq 25.$$

Find the maximum of $3x + 2y$ on D.

9. Find the maximum perimeter of a rectangle which has a diagonal less than or equal to 5 and a ratio of sides at least 4.

10. A goat is tethered to a small ring by a 39-ft rope and his movements are further restricted by a straight fence which passes 15 ft from the ring. The goat is on a slope and he has found that, the lower down the slope, the sweeter is the grass. In terms of coordinates x, y measured along the slope with the ring at $(0, 0)$ and the fence at $x = 15$, the height at a point of the slope is given by $3y - x$ in. At which point will the goat find the sweetest grass?

2.5   A Metamorphosis

The technique we have developed for solving linear programming problems has a very serious limitation. We are confined to the geometry of the plane and, if we go back to see how this arose, we see that we have considered only problems with two independent variables. In the examples of the last section, the variables were the numbers of shirts and pajamas in the first case and two kinds of fertilizer in the second. But what if the manufacturer of cotton goods had *four* items to produce—shirts, pajamas, handkerchiefs, and blouses? Obviously, we can formulate the problem in a similar way with constraints imposed by available material and working hours, and with the profit formulated as a linear objective function to be maximized. But this is as far as we can go. We can no longer draw graphs which will allow us to visualize the feasible solutions.

For such problems, we have to dispense with geometry and call on more powerful algebraic methods. We devote the remainder of this chapter to a discussion of the *simplex method* for solving linear programming problems. We will not do this in detail, but we will be able to indicate the line of argument and the techniques that are needed for a more thorough investigation.

The first idea we need indicates a method for simplifying the algebraic form of the constraints. The simplex method depends on our ability to solve simultaneous algebraic equations, and we wish to transform the conditions of *inequality* which make up the constraints into conditions of *equality* for nonnegative variables. This is easily done at the expense of introducing one further unknown quantity for each constraint involving an inequality. We have only to observe that if a, b, c are given, the condition

(9) $$ax + by \leq c$$

for real numbers x and y is *logically equivalent* to the pair of conditions

(10) $$ax + by + u = c \quad \text{and} \quad u \geq 0.$$

In other words, (9) $\Rightarrow$ (10) and (10) $\Rightarrow$ (9). Stated more simply, we introduce a new nonnegative variable to make up the difference (take up the slack) between the quantities on the left and right of the inequality (9). For example, the condition $3x - 4y \leq 4$ is equivalent to $3x - 4y + u = 4$ and $u \geq 0$, where u is introduced to make up the difference between $3x - 4y$ and 4. Similarly, $-x + 2y \geq 3$ is equivalent to $-x + 2y = 3 + u$ and $u \geq 0$.

The equivalence just described does not depend on the number of variables involved, of course. Thus, if we have variables $x_1, x_2, \ldots, x_n$, then

$$\sum_{i=1}^{n} a_i x_i \leq c \iff \left(\sum_{i=1}^{n} a_i x_i\right) + u = c \quad and \quad u \geq 0.$$

For example,

$$2x - 3y + z \leq 4$$

is equivalent to $2x - 3y + z + u = 0$ and $u \geq 0$.

Suppose that we have a problem whose first formulation in terms of inequalities involves r nonnegative variables which must satisfy s constraints in the form of linear inequalities (in the shirts, pajamas example, $r = 2$, $s = 3$). We apply the above dodge and the result is s linear *equations* in $r + s$ nonnegative unknowns. The s variables which are introduced to complete this metamorphosis from inequalities to equations are called the *slack variables*.

In order to take advantage of this, we must pay some attention to a systematic elimination method for solving simultaneous equations.

Exercises for Section 2.5

1. By introducing three slack variables taking nonnegative values, rewrite the conditions of Example 1(e) of Section 2.2 as three equations.
2. Repeat Exercise 1 using the constraints imposed in the fertilizer problem of Section 2.4.

2.6 The Elimination Method

Suppose that we have m linear equations in n unknowns. Any such set of equations can be written in the form

(11)

$$a_{11}x_1 + a_{12}x_2 + \cdots + a_{1n}x_n = b_1$$
$$a_{21}x_1 + a_{22}x_2 + \cdots + a_{2n}x_n = b_2$$
$$\vdots$$
$$a_{m1}x_1 + a_{m2}x_2 + \cdots + a_{mn}x_n = b_m.$$

For example, in Exercises 3 and 4(a) at the end of this section, we have three equations in three unknowns, and three equations in four unknowns, respectively. It is assumed that the a and b coefficients are all known and we are to find *n-tuples* of numbers $x_1, \ldots, x_n$ for which each of the m conditions in Eq. (11) is true. Every such n-tuple is a solution of the system of equations, and the set of *all* solutions is the *solution set* of the system.

It can be proved that either the solution set is empty (there are *no* solutions), or there is precisely one solution (the solution is unique), or there are infinitely many solutions.

In the case $m = n = 1$ the system reduces to one equation in one unknown, say $ax = b$, and these three alternatives are realized in the following respective cases:

$$a = 0 \quad \text{and} \quad b \neq 0, \qquad a \neq 0, \qquad a = 0 \quad \text{and} \quad b = 0.$$

The exercises at the end of this section provide some more examples.

Note that in our programming problems we will always have fewer equations than unknowns, so we may confine our attention to the case $m \leq n$.

We now wish to perform certain simple operations on our system which will change the system to a new one *having the same solution set*. When two systems of equations have the same solution set, they are said to be *equivalent*. Thus, our operations are to transform a system such as Eq. (11) into an equivalent system. By making a sequence of such transformations, we aim to simplify the array of coefficients to such an extent that we will be able to read off a solution by inspection.

We shall take it as obvious that the application of either of the following *elementary operations* transforms a system into an equivalent one:

E1. Multiply an equation by a nonzero number.
E2. Add to one equation a multiple of another equation.

The objective now is to apply operations of these two types repeatedly to obtain a system equivalent to the one we are to solve, but in a particularly simple form involving as many zero coefficients as possible.

A little exercise in applying the technique will verify that all we need to keep track of in this process is the array of numbers making up the coefficients of the system. For example, the array corresponding to the system of Eq. (11) is written

(12)

$$\begin{array}{cccc|c} a_{11} & a_{12} & \cdots & a_{1n} & b_1 \\ a_{21} & a_{22} & \cdots & a_{2n} & b_2 \\ \vdots & & & \vdots & \\ a_{m1} & a_{m2} & \cdots & a_{mn} & b_m \end{array}$$

Our objective can now be formulated as follows (recalling that we assume $m \leq n$): We apply elementary operations in such a way as to introduce m *different unit columns* into the foregoing array of a coefficients. A unit column consists of zeros with one exception, which is a 1. For example,

(13)

$$
\begin{array}{cccc|c}
0 & 5 & 1 & 0 & 9 \\
1 & -3 & 0 & 0 & 4 \\
0 & 16 & 0 & 1 & -3
\end{array}
$$

is an array of the desired type in which $m = 3$, $n = 4$.

A set of m equations in n unknowns with $m \leq n$, in which the m different unit columns appear in the array of coefficients of the unknowns, is said to be in *normal form*. Thus, the array (13) corresponds to the equations

(14)

$$
\begin{aligned}
5x_2 + x_3 \quad &= 9 \\
x_1 - 3x_2 \quad &= 4 \\
16x_2 \quad + x_4 &= -3,
\end{aligned}
$$

which are in normal form.

What advantage does a normal form have? It has the great advantage that it allows us to read off a solution (if any) by inspection. When we give the value zero to all variables associated with columns *other than* the unit columns, the coefficients on the right-hand side immediately yield the values of the other m variables which determine a solution. This is called the *basic solution* determined by the normal form. On the other hand, by assigning *any* value whatever to the variables associated with nonunit columns, we can obtain more solutions.

Applying this idea to Eqs. (14) yields the *general* solution:

$$
x_1 = 4 + 3\alpha, \qquad x_2 = \alpha, \qquad x_3 = 9 - 5\alpha, \qquad x_4 = -3 - 16\alpha.
$$

Giving α *any* numerical value results in a 4-tuple (x_1, x_2, x_3, x_4) which is a solution, and all members of the solution set can be obtained in this way. The basic solution obtained here is

$$
x_1 = 4, \qquad x_2 = 0, \qquad x_3 = 9, \qquad x_4 = -3.
$$

The following equations are in normal form:

$$
\begin{aligned}
x_1 + 2x_2 \quad + 4x_4 \quad &= 3 \\
3x_2 \quad - 2x_4 + x_5 &= 7 \\
- x_2 + x_3 \quad &= -6.
\end{aligned}
$$

By assigning values α, β to x_2 and x_4, we can write the general solution in the following form:

$$x_1 = \quad 3 - 2\alpha - 4\beta$$
$$x_2 = \qquad \alpha$$
$$x_3 = -6 + \alpha$$
$$x_4 = \qquad\qquad \beta$$
$$x_5 = \quad 7 - 3\alpha + 2\beta.$$

For a general linear programming problem of the type considered here, there will be $n - m$ parameters in a general solution.

The following points should be emphasized:

Remark 1. We have made quite a few statements without proof in this section. A precise statement of necessary and sufficient conditions for the existence of solutions involves a little more mathematical machinery than we are prepared to develop. Furthermore, all examples in this chapter will involve systems which can be reduced to normal form, although this cannot be done generally. A small amount of linear algebra will yield satisfactory answers to these interesting and important questions.

Remark 2. For any given system, there are several ways of reducing it to normal form and several possible normal forms. Each normal form yields a basic solution and a different description of the *same* solution set in the form of a general solution.

We will illustrate this discussion with an example. We will write systems of equations on the left and the corresponding arrays of coefficients on the right. This should demonstrate the advantage of working with arrays and allow the reader to do so in subsequent exercises.

Equations	Operation	Arrays				
$2x_1 + 4x_2 + ②x_3 + 8x_4 = 2$ $5x_1 \qquad - x_3 + 2x_4 = 7$ $-3x_1 + 2x_2 \qquad + 3x_4 = 2$	Select pivot	2	4	②	8	2
		5	0	−1	2	7
		−3	2	0	3	2
$x_1 + 2x_2 + x_3 + 4x_4 = 1$ $5x_1 \qquad - x_3 + 2x_4 = 7$ $-3x_1 + 2x_2 \qquad + 3x_4 = 2$	$\tfrac{1}{2} \times$ Eq. (1)	1	2	1	4	1
		5	0	−1	2	7
		−3	2	0	3	2
$x_1 + 2x_2 + x_3 + 4x_4 = 1$ $6x_1 + ②x_2 \qquad + 6x_4 = 8$ $-3x_1 + 2x_2 \qquad + 3x_4 = 2$	Eq. (2) + Eq. (1)	1	2	1	4	1
		6	②	0	6	8
		−3	2	0	3	2

FIRST CYCLE COMPLETE

Equations	Operation	Arrays

Equations	Operation	Arrays				
$x_1 + 2x_2 + x_3 + 4x_4 = 1$		1	2	1	4	1
$3x_1 + x_2 + 3x_4 = 4$	$\frac{1}{2} \times$ Eq. 2	3	1	0	3	4
$-3x_1 + 2x_2 + 3x_4 = 2$		-3	2	0	3	2
$-5x_1 + x_3 - 2x_4 = -7$	Eq. 1 + (−2)(Eq. 2)	-5	0	1	-2	-7
$3x_1 + x_2 + 3x_4 = 4$		3	1	0	3	4
$-9x_1 - ③x_4 = -6$	Eq. 3 + (−2)(Eq. 2)	-9	0	0	-③	-6

SECOND CYCLE COMPLETE

Equations	Operation	Arrays				
$-5x_1 + x_3 - 2x_4 = -7$		-5	0	1	-2	-7
$3x_1 + x_2 + 3x_4 = 4$		3	1	0	3	4
$3x_1 + x_4 = 2$	$(-\frac{1}{3}) \times$ Eq. 3.	3	0	0	1	2
$x_1 + x_3 = -3$	Eq. 1 + (2)(Eq. 3)	1	0	1	0	-3
$-6x_1 + x_2 = -2$	Eq. 2 + (−3)(Eq. 3)	-6	1	0	0	-2
$3x_1 + x_4 = 2$		3	0	0	1	2

PROCESS COMPLETE

Note that the procedure is conveniently broken down into cycles, and in each cycle a unit column is produced. The cycle consists of (a) selecting a pivot, (b) using an E1 operation to replace this pivot element by a 1 (if necessary), and (c) using operations of type E2 to produce a unit column containing the chosen pivot element.

The reader should now find the corresponding basic solution and formulate a general solution.

As a further illustration of the possible multiplicity of normal forms, consider the equations

$$2x_1 - x_2 + 2x_3 - x_4 + 3x_5 = 14$$
$$x_1 + 2x_2 + 3x_3 + x_4 = 5$$
$$x_1 - 2x_3 - 2x_5 = -10.$$

We claim that an equivalent system in normal form is

$$\frac{9}{4}x_1 + \frac{1}{2}x_2 + x_3 \qquad\qquad = \quad 2$$
$$-\frac{23}{4}x_1 + \frac{1}{2}x_2 \qquad + x_4 \qquad = -1$$
$$-\frac{11}{4}x_1 - \frac{1}{2}x_2 \qquad\qquad + x_5 = \quad 3,$$

with the basic solution $x_1 = 0$, $x_2 = 0$, $x_3 = 2$, $x_4 = -1$, $x_5 = 3$.

However, the following set is also a normal form for the original system

$$\frac{1}{11}x_2 + x_3 \qquad\qquad + \frac{9}{11}x_5 = \quad \frac{49}{11}$$
$$\frac{17}{11}x_2 \qquad + x_4 - \frac{23}{11}x_5 = -\frac{80}{11}$$
$$x_1 + \frac{2}{11}x_2 \qquad\qquad - \frac{4}{11}x_5 = -\frac{12}{11}$$

with the associated basic solution $x_1 = -\frac{12}{11}$, $x_2 = 0$, $x_3 = \frac{49}{11}$, $x_4 = -\frac{80}{11}$, $x_5 = 0$.

Exercises for Section 2.6

1. When solving two simultaneous linear equations in two unknowns, say x and y, we may represent the solution set of one equation by a straight line in the x, y coordinate plane, and similarly for the second equation. Under what geometrical conditions does the *pair* of equations have (a) no solution, (b) one solution, and (c) infinitely many solutions?

2. Write the coefficients of the following equations in tabular form, and go through one cycle of the elimination procedure (eliminate one variable only) using the circled coefficient as the pivot.

$$4x + ②y + \quad 2z = 4$$
$$-8x + \quad 4y + 16z = 7$$
$$12x - \quad 6y + \quad 3z = 4.$$

3. Reduce the following equations to a normal form and hence find their solution (there is only one in this case):

$$x + 4y + 11z = 7$$
$$2x + 8y + 16z = 8$$
$$x + 6y + 17z = 9.$$

4. Using a careful tabular method, find a normal form, a basic solution, and a general solution for each of the following systems. Use the successive pivot positions

indicated in the first two examples; (i, j) denotes the position in the ith row and jth column of an array.

(a) $x_1 + x_2 + x_3 + x_4 = 4$
    $x_1 + 2x_2 + 2x_3 + 3x_4 = 7$ Pivots $(1, 1), (2, 2), (3, 3)$
    $3x_1 + 3x_2 + 5x_3 + 4x_4 = 15$

(b) $2x_1 + x_2 + 3x_3 + 5x_4 = 6$
    $3x_1 + 2x_2 + 4x_3 + 6x_4 = 8$ Pivots $(1, 2), (2, 1), (3, 3)$
    $-x_1 + 3x_2 + 2x_3 + 7x_4 = -3$

(c) $4x + 6z - 2w = 4$
    $3x - 7y + z + 2w = 3$
         $y - 7z + w = 2$

2.7 The Simplex Method for a Special Problem

We do not propose to give a general treatment of the simplex method, but only to show how it works when applied to a problem we have already discussed. We do hope, however, that the reader will catch a glimpse of the power of the method from this illustration.

    Consider the problem posed in Section 2.1, which is described by the constraints (5):

$$x \geq 0, \quad y \geq 0$$

(15)
$$x + 2y \leq 1000$$
$$2x + y \leq 1200$$
$$2x + 3y \leq 1600$$

and the objective function to be maximized:

$$P(x, y) = x + y.$$

We first reformulate the constraints as described in Section 2.5 and obtain equivalent conditions for nonnegative x, y, u, v, w. We also incorporate the objective function in the array and write:

(16)
$$x + 2y + u = 1000$$
$$2x + y + v = 1200$$
$$2x + 3y + w = 1600$$
$$-x - y + M = 0.$$

The problem can now be stated as follows: *Find nonnegative values for x, y, u, v, w, M which satisfy Eqs. (16) and yield the largest possible value of M.* Thus, we are to examine the members of the solution set of four equations in six unknowns and find a 6-tuple in which the last component takes the largest value attained in the nonnegative members of the set.

We now go through the steps of the algorithm leading to the solution (which we have already found by other means). Observe first of all that Eqs. (16) are already in normal form. The array of coefficients is:

(17)

x	y	u	v	w	M	
1	2	1	0	0	0	1000
②	1	0	1	0	0	1200
2	3	0	0	1	0	1600
−1	−1	0	0	0	1	0

The associated basic feasible solution is

$$x_1 = 0,\ y_1 = 0,\ u_1 = 1000,\ v_1 = 1200,\ w_1 = 1600,\ M_1 = 0.$$

In the corresponding *general* solution u, v, w, M are expressed as functions of (in terms of) x and y. In particular,

$$M = x + y.$$

We now ask, can M be increased by changing x from the value $x_1 = 0$ while holding y fixed at the value $y = y_1 = 0$? Simultaneously, we must satisfy the nonnegative conditions obtained from the first three equations:

$$u = 1000 - x \geq 0.$$
$$v = 1200 - 2x \geq 0.$$
$$w = 1600 - 2x \geq 0.$$

The answer is yes. In fact we can increase x from $x_1 = 0$ as far as $x_2 = 600$. We can go no further without violating the second of the last three inequalities. The values $x_2 = 600$, $y_2 = 0$ determine a new feasible solution:

$$x_2 = 600,\ y_2 = 0,\ u_2 = 400,\ v_2 = 0,\ w_2 = 400,\ M_2 = 600.$$

On the basis of this conclusion the basic step in the algorithm is now determined as follows: Transform the system (17) to a different normal form in which the above feasible solution will be a *basic* feasible solution and x (the unknown we have just varied) is associated with the *second* unit column (because the *second* of the three inequalities was critical). This determines the pivot coefficient shown ringed in (17).

The elimination process developed in Section 2.6 now yields an equivalent system determined by the array:

	x	y	u	v	w	M	
(18)	0	$\frac{3}{2}$	1	0	0	0	400
	1	$\frac{1}{2}$	0	$\frac{1}{2}$	0	0	600
	0	2	0	-1	1	0	400
	0	$-\frac{1}{2}$	0	$\frac{1}{2}$	0	1	600

It is easily verified that the system represented does, indeed, have $x_2, y_2,$ $\dots, M_2$ as a basic feasible solution. In this case, the general solution expresses $x, u, w,$ and M in terms of y and v. In particular,

$$M = 600 + \tfrac{1}{2}y - \tfrac{1}{2}v.$$

The argument is now repeated. Can we increase M by changing y while keeping v fixed at $v = v_2 = 0$ and simultaneously satisfy the nonnegative conditions obtained from the first three equations:

$$u = 400 - \tfrac{3}{2}y \geq 0$$
$$x = 600 - \tfrac{1}{2}y \geq 0$$
$$w = 400 - 2y \geq 0?$$

Again, the answer is yes. We can increase y from $y_2 = 0$ to $y_3 = 200$ and can go no further without violating the third condition. Then the values $y_3 = 200,$ $v_3 = 0$ determine a new feasible solution:

(19) $x_3 = 500, y_3 = 200, u_3 = 100, v_3 = 0, w_3 = 0, M_3 = 700$

Apply the basic step again! Transform the system (18) to a different normal form in which the above feasible solution will be a *basic* feasible solution and y (the unknown we have just varied) is associated with the *third* unit column (because the *third* of the three inequalities was critical). This determines the pivot coefficient shown ringed in (18).

The elimination process yields a third equivalent system in normal form represented by:

	x	y	u	v	w	M	
(20)	0	0	1	$\frac{3}{2}$	$-\frac{3}{2}$	0	100
	1	0	0	$\frac{1}{4}$	$-\frac{1}{4}$	0	500
	0	1	0	$-\frac{1}{2}$	$\frac{1}{2}$	0	200
	0	0	0	$\frac{1}{4}$	$\frac{1}{4}$	1	700

What happens if we try to progress and improve the last feasible solution by another cycle of the procedure? We find that we just can't get started! On examining the "M equation" in (20),

$$M = 700 - \tfrac{1}{4}v - \tfrac{1}{4}w,$$

we see that changing x and y can now do nothing for us in our attempt to increase M. Remarkably, this implies that our solution (19) is the optimal one, yielding the largest possible value of M by choice of nonnegative x and y.

We make this assertion without proof. The prospect of a more careful analysis is left to tempt the reader to a deeper study of linear programming in the future.

If we interpret the result in the setting of the original problem, we see that we have reproduced the solution obtained in Section 2.4:

$$x_3 = x_c = 500, \qquad y_3 = y_c = 200.$$

The value $u_3 = 100$ indicates the surplus of cotton material which is left when these numbers of shirts and pajamas are made. Then v_3, w_3 have similar interpretations as surpluses of cutting time and making-up time, respectively.

Exercise for Section 2.7

1. Apply the simplex method to solve:
(a) Example 4 of Section 2.4.
(b) The fertilizer problem of Section 2.4.

2.8 Concluding Remarks

It is not difficult to see that the technique we have used above does not depend on the number of variables involved. Thus, the simplex method in its full generality can be applied (in theory) to linear programming problems with any finite number of variables. Of course, as the number of variables increases so does the number of cycles needed to reach the optimal solution and so does the amount of work in each cycle. We must expect that the amount of work needed to obtain a solution will increase with the size of the problem. However, the point is that we have indicated here a method which is not tied to the geometry of the plane, and hence to two variables, as is the case with the technique used in Section 2.4.

For problems involving three to six variables, say, the simplex method can be formalized into a convenient tabular method suitable for calculation with a desk calculator. For larger-scale problems, the algorithm lends itself very readily to programming for larger computers.

   Although we have introduced the simplex method in a manipulative, algebraic way, it has a very beautiful geometric meaning. Indeed, the geometric interpretation gives rise to the use of the word "simplex." Notice that the (x, y) values given by the basic feasible solutions of our application of the method are $(0, 0)$, $(600, 0)$, and $(500, 200)$ and that these points correspond to vertices of the region of feasible solutions (Fig. 2.4). Furthermore, this is a *chain* of vertices in the sense that we proceed from one vertex to another along a single edge of the region (a simplex) of feasible solutions. When there are n independent variables in a problem, the region of feasible solutions for a well-posed problem is located in n-space, it is convex, and is bounded by so-called "hyperplanes." Such a region is called a *simplex*, and the simplex method, as outlined above, determines an algorithm which picks out a chain of vertices (one determined by each basic feasible solution) which terminates at a vertex where the objective function takes its greatest value.

Preliminary Exercises

1. If L is a straight line and distinct points with coordinates (x_1, y_1), (x_2, y_2) (referred to rectangular Cartesian axes) lie on L, then the *slope* of L is defined as

$$m = \frac{y_2 - y_1}{x_2 - x_1} \qquad \text{(if } x_2 \neq x_1\text{)}.$$

We assign the symbol ∞ to m if $x_2 = x_1$.
(a) If θ is the (counterclockwise) angle made by L with the positive direction along the first coordinate axis, $(0 \leq \theta < 180°)$, show that $m = \tan \theta$.
(b) Calculate the slope of the line joining the pairs of points

$$(1, 2) \quad \text{and} \quad (2, 4), \qquad (-3, 5) \quad \text{and} \quad (-2, -1), \qquad (13, 2) \quad \text{and} \quad (-4, 2)$$
$$(1, -1) \quad \text{and} \quad (1, 3), \qquad (3, 9) \quad \text{and} \quad (0, 0).$$

(c) Let the point P with coordinates (x, y) lie on the line joining P_1 and P_2 and assume $m \neq \infty$. Show that a condition satisfied by the coordinates of all such points P (and no others) is

$$\frac{y - y_1}{x - x_1} = \frac{y_2 - y_1}{x_2 - x_1}$$

and hence that the equation of L may be written

$$y - y_1 = m(x - x_1).$$

(d) Find equations of the lines through $(2, -1)$, $(4, 0)$, and $(2, 1)$ with slopes 6, $-1/2$, and ∞, respectively.

(e) Show that $y = mx + d$ is the equation of a line with slope m passing through $(0, d)$.

(f) Find the equations of the six lines with slopes $6, -5, 0$ through $(0, 1)$, $(0, -3)$.

(g) Sketch the straight lines whose equations are

$$y = 2x, \qquad y + x = 2, \qquad 2y = 6x - 3.$$

2. Describe the two families of straight lines obtained as follows:
   (a) Lines with equations $y = mx - 1$, where m can take any real value or ∞.
   (b) Lines with equations $y + x = k$, where k can take any real value.

3. Sketch the graph of the line whose equation is $2x + 4y = 6$, and indicate on the same sketch the points with coordinates $(0, 0), (3, -1), (4, 2), (-1, -3), (5, -1), (-1, 0), (1, 2)$. Note which of these points have the properties $2x + 4y <, =,$ or > 6.

4. Find the solution sets of (solve simultaneously) the following pairs of conditions, and sketch graphs showing the solution set of each condition separately:

$$\left.\begin{array}{l} x \qquad\ = 2 \\ 3x - 4y = 6 \end{array}\right\} \qquad \left.\begin{array}{l} 3x + 4y = 2 \\ 2x + 3y = 1 \end{array}\right\} \qquad \left.\begin{array}{l} 2x - 2y = \ \ 3 \\ x + 4y = -1 \end{array}\right\} \qquad \left.\begin{array}{l} x + 2y = \ \ 1 \\ 3x + 6y = -1 \end{array}\right\}$$

5. The relation $a > b$, where a, b are real numbers, means that the number $a - b$ is positive. Prove:
   (a) If $a > b$ and c is any real number, then $a + c > b + c$.
   (b) If $a > b$ and c is any positive real number, then $ac > bc$.
   Hence prove:
   (c) If $a > b$ and $c > d$, then $a + c > b + d$.
   (d) If x, y are real numbers for which $8 \geq 3x + 4y$, then $y \leq -\frac{3}{4}x + 2$.

6. Sketch graphs of the solution sets of
   (a) $y = x^2$
   (b) $(y - 3)^2 + x^2 = 4$
   (c) $y + 2 = x^2 - 4x + 5$

CHAPTER THREE

GROWTH OF POPULATIONS

3.1 Introduction

In this chapter we consider some mathematical models that can be used to describe variations in population of various animal species. In practice, observations of animal populations indicate either a roughly constant size from year to year, or a fluctuating population size which varies about some intermediate value. The variations about this intermediate, or equilibrium, value may be spasmodic in response to unpredictable phenomena such as weather or disease, or the variations may show a clear trend with time, e.g., exponential growth or decay, or periodic fluctuations. It is our objective to show how some simple models can be used to indicate the variations in population that arise from certain obviously important effects taken in isolation. As indicated in Chapter 1, the construction of a mathematical model involves essentially two parts. First is the idealization of the phenomenon; this may involve simplifying its form and neglecting effects which are thought to be insignificant. Second is the imposition of certain mathematical laws based on physical intuition or, better still, on observation of similar phenomena.

For the problems we will consider, decisions regarding the most important factors to be accounted for seem to be more difficult than those in problems of mechanics, for example. Animal life can be subject to so many important and unpredictable effects such as weather, other animal species, and disease that the mathematical models are of a speculative nature. However, they do show the immediate effect of relevant factors acting alone or in small combinations. They may also be useful in explaining and predicting qualitatively the variations in population from one generation to another. Possible weaknesses of the models are discussed again in Section 3.11.

Consider first an animal species whose population is sampled at annual intervals. Suppose that, *on the average,* for each member of the population alive at a given sampling, R members of the population are alive one year later. This quantity (or *parameter*), R, is of prime importance in any study of variations in population size. In the first half of this chapter, we study the effects of various hypotheses concerning R and exclude all other possible influences on population size.

It should also be remarked that our definition of R is carefully worded and the value of R is determined by some amalgam of birth *and* death rates for the species. Indeed, Eq. (1) below may be a clearer form of definition for those accustomed to mathematical symbolism.

The choice of annual intervals is purely for convenience of expression. The period used in any real-world situation would probably depend primarily on the life-span or reproductive cycle of the species, which could include bacteria, insects, or mammals. Indeed, some species of insects are more readily observable under laboratory conditions.* One further detail: the *number* of animals will generally refer to some unit determined by the environment in which observations are made, whether in the field or in the laboratory. We cannot pursue the question of the appropriate unit at this stage, since this varies from one case study to another. We are more interested in those features common to many different situations.

As part of a preliminary, very simple model, we suppose that for a number of years R has the same numerical value in each year, and that we begin to take an interest in season number zero when there were X_0 animals. Suppose further that there were $X_1, X_2, \ldots, X_n$ animals in years one, two, $\ldots, n$. With our hypothesis on R, we can say that

$$\text{(Population in year } n) = R \times (\text{Population in year } n - 1),$$

or

(1) $$X_n = RX_{n-1}$$

for $n = 1, 2, 3, \ldots$. It is claimed that we now have a model that will allow us to predict the population in year n given R and X_0.

Indeed, we have $X_{n-1} = RX_{n-2}$, and so

$$X_n = R(RX_{n-2}) = R^2 X_{n-2}.$$

Then, if $n - 2 \neq 0$, put $X_{n-2} = RX_{n-3}$. Repeating this process, we finally

*See, for example, Clark et al., (1967).

arrive at

(2) $X_n = R^n X_0.$

Now if $R = 1$, the population remains constant. If $R < 1$, then the population decreases year by year, and the ultimate extinction of the population is predicted. If $R > 1$, the population will increase indefinitely. The variation in population predicted by this model is indicated in Fig. 3.1 for three different values of R. (Note that here, and subsequently in this chapter, some graphs consist of a set of discrete points. The points are joined by segments of straight lines only to make the graphs easier to read.)

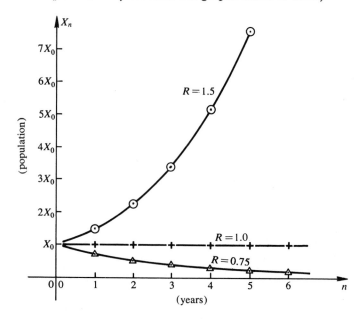

Figure 3.1 Variation in population with constant rate of change.

3.2 Rate of Growth Dependent on Population

The reader will have no difficulty finding inadequacies in the above model! It is physically unrealistic to allow R to remain constant and different from 1 for very long. If the species is to survive and is not to dominate the world, there must be some "resilience" in the rate of growth R of the population. For example, after a low breeding season and subsequent drop in population, it must be possible for R to increase again to bring the population of the species back to normal. This suggests—and it is very plausible—that if the population is reduced, then certain natural causes are called into play which cause R to increase. These may include greater availability of foodstuffs, for example. On the other hand, if the population X gets *too* low, there may be no hope for recovery and the species may be headed for extinction, perhaps because females do not find mates.* In this case, for very low X, we may expect R to decrease as X decreases. Finally, if the population increases beyond a "normal" density, then the effects of overcrowding, in one form or another, may give rise to low rates of reproduction, R.

All of these effects are reflected in Fig. 3.2 which shows a sketch of a possible graph of R as a function of X.

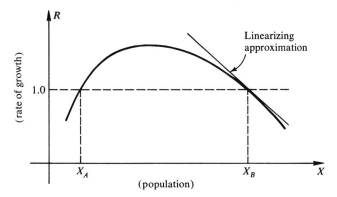

Figure 3.2 Rate of growth, R, as a function of population, X.

A population level giving rise to a rate of growth $R = 1$ is called an *equilibrium* level. If the population is at an equilibrium level in one year, then the model predicts that the population will be the same in the following year. The populations (or population densities) represented in Fig. 3.2 by X_A and X_B are equilibrium levels. They are, however, of rather different character. The population X_B represents a *stable* equilibrium value because small variations in population about X_B give rise to changes in R which tend to return subsequent populations toward the equilibrium value. For example, a small *decrease* in X below the level X_B causes an *increase* in R and, hence, a subsequent increase in X which may return the population to somewhere near X_B. In contrast to this, if the population changes by a small amount from the equilibrium value X_A, the resulting change in R causes a bigger subsequent change in population in the same direction. For this reason, X_A is called a position of *unstable* equilibrium.

Since populations are always subject to small variations, it is clear that positions of stable equilibrium will generally be the most significant.

What we are indicating here, in a very intuitive way, is the idea that R, the annual growth rate, should be thought of as a *function* of X, the population. That is, to each given value of X (of physical interest) we would like

---

*By 1972, it was believed that the world population of certain species of whale was dangerously near to this "reproduction threshold."

to assign a unique corresponding value of R. Obviously, this is not possible in real life, but we can use it as an idealization leading to a mathematical model which should, at least, be better than that of section 3.1! Once the relationship is expressed in this way, we see that—since assigning the *same* value to R for *every* X (as in Section 3.1) is just one choice of function—what we are looking for is merely a more complicated—and realistic—function.

Exercises for Section 3.2

1. The following table gives values of population density X_n in year n for $n = 0, 1, 2, 3, 4, 5, 10,$ and 15 and assuming four different initial values for X_0, namely, 1.5, 2.5, 7.5, and 8.5. The figures are based on the R-X graph sketched (Fig. 3.3). What are the equilibrium population densities? Sketch four graphs of X_n versus n for $n = 0, 1, \ldots, 5$, and examine the results carefully in the light of the definition of stable and unstable equilibrium levels.

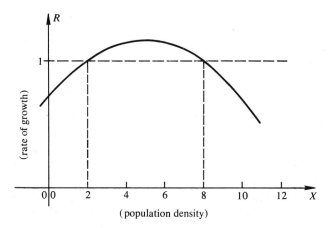

Figure 3.3

			X_n	
n	Case 1	Case 2	Case 3	Case 4
0	1.5	2.5	7.5	8.5
1	1.411	2.625	7.875	7.998
2	1.312	2.785	7.980	8.000
3	1.202	2.993	7.997	8.000
4	1.084	3.263	8.000	8.000
5	0.959	3.618	8.000	8.000
10	0.373	7.156	8.000	8.000
15	0.089	8.000	8.000	8.000

2. (a) Examine the graph (Fig. 3.4) of population density X versus time, t. Is it a plausible relation on physical grounds?

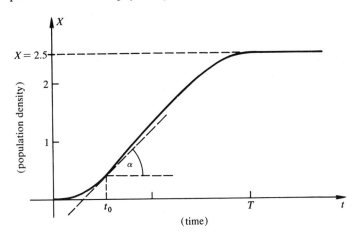

Figure 3.4

(b) Recall that the slope of the graph at some point t is given by the slope of the tangent at that value of t (for example, $\tan \alpha$ is the slope at t_0 in the figure). This slope is the *rate of change* of X with t; call it R. Sketch very roughly a graph of R (on a vertical axis) versus X (on the horizontal axis). Compare it with Fig. 3.2. [For those familiar with trigonometric functions, the above graph is that of the function

$$X(t) = \frac{5}{4}\left(1 - \cos\frac{\pi t}{T}\right) \qquad \text{for } 0 \leq t \leq T.]$$

3. Verify that the graph of

$$R = -\frac{1}{8X}(X^2 - 21X + 20)$$

has the main features of Fig. 3.2. What are the approximate equilibrium values of X?

3.3 The Linearized Problem

It turns out that, if we try to retain all the main features of Fig. 3.2, our function is too complicated for detailed analysis. We can, however, discuss an important special problem with a relatively simple function. We have in mind the investigation of *small* variations in population about a position of *stable* equilibrium. For this purpose, the properties of R in the neighborhood of X_B (in Fig. 3.2) are all that interest us. These properties can be represented quite well if we replace the curve by an approximating straight line—the tangent to the curve at X_B, for example. This process is known as *linearization* of the phenomenon, and it must be borne in mind that its use is confined to small variations in population (relative to X_B) about X_B.

Notice that in these two sections we have approached one model of the real-life situation (R a rather complicated function of X), rejected it on the grounds that it will lead to mathematics which is too difficult for us, and replaced it by a simpler model which should, nevertheless, retain some interesting features of the original problem. This is not an unusual situation in applied mathematics.

To return to our linearized problem, we now consider the case in which the graph of R versus X is a straight line. Thus, for some constants a and b, we shall have

$$(3) \qquad\qquad R = aX + b.$$

Since we are now focusing attention on variations in X about X_B, we introduce a new variable x defined by $x = X - X_B$ and consider only those cases where x/X_B is small compared to 1. If we introduce a new constant $b_1 = b + aX_B$, the equation for R as a function of x becomes

$$R = ax + b_1.$$

Notice that x may be viewed as the increase in population over X_B provided a decrease is interpreted as a negative increase. We can immediately determine one of these constants since we require that $R = 1$ when $x = 0$ (when $X = X_B$); thus $1 = 0 + b_1 = b_1$. A second feature of the problem can be built into this equation if we note that, since X_B is a position of *stable* equilibrium, the slope of the line is negative. We therefore write $a = -m$, where $m \geq 0$, and the equation has the final form

$$(4) \qquad\qquad R = 1 - mx.$$

We shall not allocate a numerical value for m, but retain this as a parameter for the problem. Thus, we will be able to formulate a solution for X_n in terms of X_0, X_B, and m. Particular solutions can then be obtained by substitution of appropriate numerical values for these three quantities.

A final word on the description of x. It should be borne in mind that X may be thought of in two ways: either as the population or as the population *density*. For example, X may be the number of animals per cubic centimeter in the case of bacteria, per square yard in the case of insects, or per acre in the case of mammals. It is convenient, therefore, to suppose that a measure of X is chosen for which X_B is approximately 1. If this is done, then x itself is restricted to have numerical values very much less than one.

3.4 Approximation to the Solution of the Linearized Problem

Our basic equation is now:

(Population in year n) =
    (Rate of increase in year $n-1$) (Population in year $n-1$)

or

$$X_n = R_{n-1}X_{n-1} \qquad (\text{for } n = 1, 2, 3, \ldots).$$

If we write $X_n = X_B + x_n$, $X_{n-1} = X_B + x_{n-1}$ and, using Eq. (4), $R_{n-1} = R(x_{n-1}) = 1 - mx_{n-1}$ for the rate of increase in year $n-1$, we obtain

(5) $$X_B + x_n = (1 - mx_{n-1})(X_B + x_{n-1}).$$

Realize that, in this equation, X_B and m are viewed as known quantities and we are trying, first of all, to express x_n in terms of x_{n-1} and then to obtain x_n in terms of $x_0 = X_0 - X_B$, as was done in obtaining Eq. (2) from Eq. (1).

Rearranging Eq. (5), we get

(6) $$x_n = (1 - mX_B)x_{n-1} - mx_{n-1}^2 \qquad (\text{for } n = 1, 2, \ldots).$$

From a computational point of view, this may be thought of as the solution. Given X_0 and hence x_0, we simply put $n = 1$ and substitute numerical values for m, X_B, and x_0 to obtain x_1. Then we set $n = 2$, and using the previously obtained x_1, we substitute and calculate x_2. Then we calculate x_3 from x_2, and so on. However, Eq. (6) is still too difficult for us to be able to obtain a precise formula giving x_n in terms of x_0. But observe that our model already restricts us to values of x_{n-1} which are much less than 1; therefore, in Eq. (6), the term mx_{n-1}^2 is likely to be much smaller than the other terms. If we introduce a new sequence, $\xi_0, \xi_1, \xi_2, \ldots$, where $\xi_0 = x_0$ and

(7) $$\xi_n = (1 - mX_B)\xi_{n-1}$$

then we expect to get a good approximation to the solution of Eq. (6) provided the x's and ξ's are small compared to 1. We shall discuss just how good this approximation is in the next section.

If we now investigate Eq. (7) and write $\alpha = 1 - mX_B$, we have simply

$$\xi_n = \alpha\xi_{n-1}$$

for $n = 1, 2, 3, \ldots$ and $\xi_0 = x_0 = X_0 - X_B$. Mathematically speaking, we are back to the very simple problem discussed in Section 3.1 which we know to have the solution

$$\xi_n = \alpha^n \xi_0, \qquad n = 0, 1, 2, \ldots.$$

The essential difference is that, although α is still constant for any given problem, we do not know *a priori* whether it will be positive or negative. Since $\alpha = 1 - mX_B$, this will depend on the magnitudes of m and X_B. It is clear that if $\alpha < 0$, then the signs of $\xi_0, \xi_1, \xi_2, \ldots$ will alternate and so predict a population which oscillates annually above and below the level X_B. The oscillations will increase or decrease in magnitude accordingly as $\alpha < -1$ or $-1 < \alpha < 0$. For positive values of α, the behavior is like that of the problem described in Section 3.1. The value of α may be said to *regulate* the variation in population. In the following table, we summarize the behavior to be expected of the numbers $\xi_0, \xi_1, \xi_2, \ldots$ for various ranges of α:

$\alpha < -1$	Oscillatory of increasing magnitude
$\alpha = -1$	Oscillatory of constant magnitude
$-1 < \alpha < 0$	Oscillatory of decreasing magnitude
$\alpha = 0$	Constant and equal to zero from ξ_1 onward
$0 < \alpha < 1$	Decreasing monotonically to zero
$\alpha = 1$	Constant and equal to ξ_0
$1 < \alpha$	Increasing monotonically.

Since we defined X_B to be a position of stable equilibrium with the understanding that $m \geq 0$, it follows that, since $\alpha = 1 - mX_B$, we must have $\alpha < 1$. Thus, the last two cases in the table cannot occur. However, with this information we could very easily discuss a position of unstable equilibrium characterized by a line with slope $m < 0$.

It should also be noted that, even though we called X_B a position of stable equilibrium, it is possible that the case $\alpha < -1$ can arise when the situation is *unstable*, in the sense that small variations about X_B will subsequently be magnified. In either of the cases $\alpha > 1$ or $\alpha < -1$ (i.e., $|\alpha| > 1$), the prediction is that the magnitude of the variations would eventually increase beyond the range of validity of the model. In other words, either the linearizing assumption or the substitution of Eq. (7) for Eq. (6) would be invalidated.

A formal definition of stability for the approximating relation Eq. (7) might run as follows: The relation is *stable*, *neutrally stable*, or *unstable* according as $|\alpha| < 1$, $|\alpha| = 1$, or $|\alpha| > 1$. Thus, for stability of Eq. (7) we require $|1 - mX_B| < 1$ which is equivalent to $0 < mX_B < 2$. Thus, in addition to having m and X_B positive, we require $X_B < 2/m$. Stability of Eq. (7) does not imply stability for Eq. (6), of course, but we anticipate that $X_B < 2/m$ would give some indication that solutions of Eq. (6) would be strictly decreasing in absolute value.

We illustrate the methods discussed in this section with a numerical example. Consider a population with an equilibrium level $X_B = 1$. Suppose it is known from earlier observations that $m = \frac{1}{2}$ for this species. In a particular year, which we make year number 0, a population density of $X_0 = 1.02$ is observed. What population levels does the model of this section predict for subsequent years?

Observe first of all that $\alpha = 1 - mX_B = \frac{1}{2}$ and that $x_0 = X_0 - X_B = 0.02$. We now calculate values for $x_1, x_2, \ldots$ successively from Eq. (6) by putting $n = 1$, then $n = 2$, and so on. The results for four subsequent years are tabulated below with an accuracy of six decimal places.

n	x_n	ξ_n
0	0.02	0.02
1	0.009 8	0.01
2	0.004 852	0.005
3	0.002 414	0.002 5
4	0.001 204	0.001 25

If we calculate the first members of the ξ sequence given by Eq. (7) and $\xi_0 = x_0 = 0.02$, we obtain the numbers in the last column with a much easier calculation. Since x_0 is, indeed, considerably smaller than X_B, and since the x sequence is decreasing, the ξ sequence obviously provides a good approximation to the x sequence. The ξ sequence has the further advantage that we can calculate any particular member, say ξ_4, very easily without first calculating the intermediate values, ξ_1, ξ_2, and ξ_3.

Suppose that we modify the problem by assuming a different slope for the R-x graph, namely $m = 1$. Then $\alpha = 0$ and the behavior of the sequences is quite different, as the following table illustrates. However, the point at issue is that the ξ sequence still remains "close" to the x sequence. This is the idea we will investigate and make more precise in the next section.

n	x_n	ξ_n
0	0.02	0.02
1	−0.000 400	0
2	0.000 000	0
3	0.000 000	0
4	0.000 000	0

Exercises for Section 3.4

1. In a certain region, a normal population of adult butterflies is given by $X_B = 1$, and it is known from earlier observations of the species that $m = 1.5$. After a "good" year, the population is given by $X_0 = 1.05$.

    (a) Calculate the variation in population predicted by Eq. (6) and Eq. (7) for the four subsequent years.

    (b) Repeat part (a) with $m = \frac{1}{4}$.

Results should be displayed neatly in tabular form. Do not calculate with more than three significant figures.

3.5   Errors in the Approximate Solution

For the biologist, it may be sufficient to say that the solutions of Eq. (7) are practically the same as those of Eq. (6) for small x. For the mathematician, it is inexcusable to pass by this imprecise statement if it is within his power to make it more precise. Given only two techniques, it is within our power to give some estimate of how close ξ_n will be to x_n. These techniques are, first, the manipulation of absolute values of real numbers and inequalities and, second, proof by induction.*

First a word about absolute values. In the usual way, any real number corresponds graphically to a point on a line. A point 0 on this *real line* is fixed to correspond to the number zero, positive numbers correspond to points on one side of 0, and negative numbers to points on the other side of 0. We subsequently identify the numbers with the points on the line. If x is a real number, then $|x|$, the absolute value of x, denotes the distance from x to 0. We require that *distance* be always a nonnegative real number; thus there are two points whose distance from 0 is $|x|$, namely, x and $-x$. We may *define*

$$|x| = \max(x, -x).$$

From this it is not too difficult to prove that, for any two real numbers x and y,

(8) $$|x + y| \leq |x| + |y| \quad \text{and} \quad |xy| = |x\|y|.$$

We shall be particularly interested in the distance between real numbers x and y. By looking at the real line, it is clear that the measure of distance we need is either $x - y$ or $y - x$, whichever is positive. That is, $|x - y|$ can be interpreted as the distance between x and y. Our problem is to find positive

*See, for example, Mostow, Sampson, and Meyer, *Fundamentals of Linear Algebra*, McGraw Hill, 1963. Also Preliminary Exercises 3.14, 3.15, and 3.16.

numbers k_n such that the distance between ξ_n and x_n will not exceed k_n, i.e., for which

$$|\xi_n - x_n| \leq k_n \qquad (\text{for } n = 0, 1, 2, \ldots).$$

Suppose now that we subtract Eq. (6) from Eq. (7) and write $\alpha = 1 - mX_B$, as before. Then

(9) $$\xi_n - x_n = \alpha(\xi_{n-1} - x_{n-1}) + mx_{n-1}^2.$$

We have assumed that $\xi_0 - x_0 = 0$ and we claim* that it can now be proved by induction that, for $n = 1, 2, 3, \ldots$,

(10) $$\xi_n - x_n = m(x_{n-1}^2 + \alpha x_{n-2}^2 + \cdots + \alpha^{n-2}x_1^2 + \alpha^{n-1}x_0^2).$$

The first step in the proof is to verify the result for $n = 1$. This is done simply by putting $n = 1$ in Eq. (9) and noting that $\xi_0 - x_0 = 0$. Thus,

$$\xi_1 - x_1 = mx_0^2$$

which is the special case of Eq. (10) for $n = 1$.

The second step in the proof consists in assuming Eq. (10) to be true for any particular integer, say v, and on this hypothesis we are to prove that Eq. (10) is true for the next integer, $v + 1$. Now we know from Eq. (6) and (7) that

$$\xi_{v+1} - x_{v+1} = \alpha(\xi_v - x_v) + mx_v^2$$

and, using the induction hypothesis, we may substitute for $\xi_v - x_v$ from Eq. (10) in the right side of this equation. Rearranging this expression, we arrive at

$$\xi_{v+1} - x_{v+1} = m(x_v^2 + \alpha x_{v-1}^2 + \cdots + \alpha^{v-1}x_1^2 + \alpha^v x_0^2),$$

which is what we wanted to prove.

We have agreed that our analysis is valid only for small values of x_n. Let us be more precise about this and say that, for some known real number $\epsilon > 0$, we allow only those values of x_n for which $x_n^2 \leq \epsilon$. Thus

$$|\xi_n - x_n| \leq m(x_{n-1}^2 + |\alpha|x_{n-2}^2 + \cdots + |\alpha|^{n-2}x_1^2 + |\alpha|^{n-1}x_0^2)$$

and since $x_{n-1}^2 \leq \epsilon, \ldots, x_0^2 \leq \epsilon$,

$$|\xi_n - x_n| \leq m(1 + |\alpha| + |\alpha|^2 + \cdots + |\alpha|^{n-1})\epsilon.$$

$$= \left(\frac{1 - |\alpha|^n}{1 - |\alpha|}\right)m\epsilon$$

*As indicated in Preliminary Exercise 14, the statement of Eq. (10) was first guessed after Eq. (9) was used to compute $\xi_1 - x_1$, $\xi_2 - x_2$, and $\xi_3 - x_3$ explicitly.

provided $|\alpha| \neq 1$. At the last step, we have summed the finite geometric progression* whose first term is 1 and whose common ratio is $|\alpha|$.

If we define

(11)
$$k_n = \left(\frac{1 - |\alpha|^n}{1 - |\alpha|}\right) m\epsilon$$

then $|\xi_n - x_n| \leq k_n$ and we have the desired result. The distance between x_n and ξ_n cannot exceed k_n, and if x_n is small ($x_n^2 \leq \epsilon$) then, since the factor ϵ appears in the definition of k_n, we can see at once that there is a wide class of problems in which k_n is also small, but in a way we can now measure.

Let us illustrate these results with the numerical example in Section 3.4. We again suppose that the equilibrium population is given by $X_B = 1$ and the slope of the (linearized) X-R graph is $m = \frac{1}{2}$. Then $\alpha = 1 - mX_B = \frac{1}{2}$. We then have $X_0 = 1.02$ and $\xi_0 = x_0 = 0.02$. The values of ξ_n for $n = 1, 2, 3, 4$ are calculated from $\xi_n = \alpha\xi_{n-1}$ and are tabulated below along with the numbers k_n of Eq. (11). We take for ϵ the number $x_0^2 = 0.0004$.

n	0	1	2	3	4
ξ_n	0.02	0.01	0.005	0.002 5	0.001 25
k_n	0	0.000 2	0.000 3	0.000 35	0.000 375

We have seen that the ξ_n's are very easy to calculate, and the second row tells us that they cannot differ from the x_n's by more than the corresponding k_n. As one might expect, the usefulness of the k_n's decreases as n increases. However, if we start with a smaller x_0, this will admit a smaller ϵ, and hence closer bounds, k_n, for the deviation of ξ_n from x_n.

With just a little fortitude, the x_n's are calculated from Eq. (6). They are reproduced here for the purpose of comparison:

n	0	1	2	3	4
x_n	0.02	0.009 8	0.004 852	0.002 414	0.001 204

The validity of the statement $|\xi_n - x_n| \leq k_n$ is easily verified for each n.

Remember, however, that the purpose of the exercise is to show that computation of the x's, which is relatively complicated, can be avoided. If we merely compute the ξ's, we have the benefit of the above *error analysis* which allows us to decide whether this results in any serious numerical error.

*See Preliminary Exercises 3.5, 3.6, and 3.7.

It may be argued, with some justice, that error bounds, k_n, which depend on n are too cumbersome to be very useful. In practice, bounds which are independent of n, as illustrated in Exercise 1, are likely to be more useful.

Exercises for Section 3.5

1. All notation is defined in Section 3.5. Assume throughout that $m > 0$.
   (a) Prove that, if $|\alpha| < 1$, then for $n = 1, 2, \ldots$,

$$|\xi_n - x_n| < \frac{m\epsilon}{1 - |\alpha|}$$

   (This bound has the advantage that it is easier to compute than Eq. (11) and is the same for *every n*.)
   (b) Prove that, if $0 < \alpha < 1$, then $\xi_n > x_n$ and $\xi_n - x_n < \epsilon/X_B$.
2. (a) If $m = \frac{1}{10}$, $X_B = 1$, and $X_0 = 1.02$, show that the ξ sequence is decreasing and calculate the first five terms.
   (b) Taking $\epsilon = 0.0004$, use Exercise 1 to show that, for every n, ξ_n and x_n differ by less than 0.00005.
3. Calculate bounds for $|x_n - \xi_n|$ in both cases described in Exercise 1 of Section 3.4.
4. Prove by induction that, for any real numbers $a_1, a_2, \ldots, a_n$,

$$|a_1 + a_2 + \cdots + a_n| \leq |a_1| + |a_2| + \cdots + |a_n|.$$

3.6 Interaction of Predator-Prey Species

We shall now attempt to set up models describing the effects of two species on one another. The two species are related by the fact that one is a source of food for the other. The first assumption we make is that, in the absence of the predator, the population of the prey species is governed by the linearized equilibrium model formulated in Section 3.3 Thus, we immediately confine our attention to small variations in population of the prey about a stable equilibrium level.

The degree of dependence of the predator species on the prey may vary from one situation to another. In some cases, the prey may be the principal or only source of food for the predator, and so the predator population may be controlled by that of the prey. We shall consider this in the next section. At the other extreme, the predator may devour the prey regularly but only as a part of its diet, or possibly as part of the diet which can easily be replaced by other sources of food. The latter is the simpler case, and we shall try to set up a mathematical model for it in this section.

More specifically, suppose that the predator species arrives in the region of the prey in year zero when the population of the prey is at its equilibrium value X_B. Thus $X_0 = X_B$ and $x_0 = 0$. Suppose also that in each subsequent year the same number, N, of prey is taken by the predator. Again, it may be more convenient to measure X as a population density; N is then the number of prey taken annually per unit area or volume. The equation governing the population of prey is now obtained immediately if we subtract N from the right-hand side of the basic equation of Section 3.4. Thus,

$$X_n = R_{n-1}X_{n-1} - N \qquad \text{(for } n = 1, 2, 3, \ldots\text{)}.$$

We then rearrange and find that (as in obtaining Eq. (6) from Eq. (5))

(12) $$x_n = \alpha x_{n-1} - m x_{n-1}^2 - N \qquad (n = 1, 2, 3, \ldots)$$

where $\alpha = 1 - mX_B$ as before. This relation gives the increment in density, x_n, above the equilibrium prey population X_B (in the absence of predators) in terms of the increment x_{n-1} at year $n - 1$.

In this model we are not concerned with the population of predators. In fact, we are assuming that the population of predators does *not* vary in such a way that the number of prey taken in any one year will be significantly different from that taken in any other year. Note also that N must be small in the same sense that x_n is small compared to X_B.

Again, we try to find an approximating sequence $\xi_0, \xi_1, \xi_2, \ldots$ for $x_0, x_1, x_2, \ldots$, and guided by the discussion of Section 3.4 we consider the relation

(13) $$\xi_n = \alpha \xi_{n-1} - N \qquad (n = 1, 2, 3, \ldots).$$

with $x_0 = \xi_0$. Let us first examine the initial terms of the ξ sequence. We have

$$\xi_1 = \alpha \xi_0 - N$$
$$\xi_2 = \alpha \xi_1 - N = \alpha(\alpha \xi_0 - N) - N$$
$$= \alpha^2 \xi_0 - (1 + \alpha)N$$
$$\xi_3 = \alpha \xi_2 - N = \alpha(\alpha^2 \xi_0 - (1 + \alpha)N) - N$$
$$= \alpha^3 \xi_0 - (1 + \alpha + \alpha^2)N.$$

A pattern is already emerging and we guess that, for $n = 1, 2, 3, \ldots$,

$$\xi_n = \alpha^n \xi_0 - (1 + \alpha + \alpha^2 + \cdots + \alpha^{n-1})N.$$

We then use induction to prove very easily that this is true. Using the expression for the sum of a geometric progression and observing that $\xi_0 = x_0 = 0$, we have (provided $\alpha \neq 1$)

$$(14) \qquad \xi_n = -\left(\frac{1 - \alpha^n}{1 - \alpha}\right) N$$

We must again ask: What relation does ξ_n bear to x_n? This is very easily answered now because, if Eq. (12) is subtracted from Eq. (13), we get precisely the relation in (9) which has already been studied. We therefore have $|\xi_n - x_n| \leq k_n$, where k_n is given by Eq. (11).

The behavior of the sequence ξ_n for two values of α, with $N = 0.05$, is illustrated in Fig. 3.5. Note that, if we imagine X_B to be the same in both cases, then $\alpha = -\frac{1}{2}$ corresponds to a higher value of m (determining the slope of the linearizing approximation of Fig. 3.2) than that for $\alpha = \frac{1}{2}$.

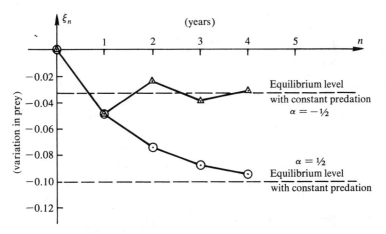

Figure 3.5 Prey populations with constant predation.

An interesting feature of this sketch is the suggestion that, after a number of years, the population of the prey "flattens off" to a new equilibrium value. It is not difficult to see this from Eq. (14). If $|\alpha| < 1$, then $|\alpha|^n$ decreases and approaches zero as n increases. It follows that, as n increases, ξ_n approaches the value $-N/(1 - \alpha)$. Thus, to the extent that ξ_n approximates x_n, we can expect a new equilibrium population of approximately

$$(15) \qquad X_B - \frac{N}{1 - \alpha} = X_B - \frac{N}{mX_B}.$$

To obtain the second expression, we substitute $\alpha = 1 - mX_B$ in the first.

We can arrive at an expression for the new equilibrium population in another way. First we use Eq. (4) to obtain $R_{n-1} = 1 - mx_{n-1}$ and write our basic equation in the form

$$X_n = (1 - mx_{n-1})X_{n-1} - N.$$

We then set $x_{n-1} = X_{n-1} - X_B$ to obtain

(16) $$X_n = (1 - mX_{n-1} + mX_B)X_{n-1} - N.$$

If X_s is the new equilibrium population with the constant level of predation, N, and if at any time we have $X_{n-1} = X_s$, then the definition of equilibrium population implies that we must also have $X_n = X_s$. Substituting $X_n = X_{n-1} = X_s$ in the last equation and rearranging, we have

$$mX_s^2 - (mX_B)X_s + N = 0.$$

This is a quadratic equation for X_s, and it follows that

(17) $$X_s = \frac{1}{2}\left\{X_B \pm X_B \sqrt{1 - \frac{4N}{mX_B^2}}\right\}.$$

Which sign should we choose? Or do both signs yield sensible solutions? To decide this question, we have only to remember that X_s must be "close" to X_B because we are still working with the assumption of *linear* variation of R with X. Thus the results can be valid only if predicted populations are close to the equilibrium population X_B. Now it is clear that if we take the plus sign in Eq. (17), then if N/X_B is small enough, X_s will be close to X_B. If we take the minus sign, the result is a positive value for $X_s \ll X_B$, which would immediately take us out of the range of validity of the linearizing approximation. Thus we have

(18) $$X_s = \frac{1}{2}\left\{X_B + X_B\left(1 - \frac{4N}{mX_B^2}\right)^{1/2}\right\}.$$

How does this compare with Eq. (15), the equilibrium population with predation predicted by the first analysis? The two results can be seen to be consistent if we note that N/X_B must be very much less than 1. Also $\alpha \neq 1$ implies $m \neq 0$, therefore we may take it that $4N/(mX_B)$ is much less than 1. Then, using the binomial expansion (see Appendix 1 of this chapter), we obtain

$$\left(1 - \frac{4N}{mX_B^2}\right)^{1/2} \simeq 1 - \frac{2N}{mX_B^2}.$$

Substituting this into (18), we obtain

$$X_s \simeq X_B - \frac{N}{mX_B},$$

which agrees with result (15).

We may now ask whether, in problems of this kind, the equilibrium population *with* predation X_s is not more significant than that without, X_B. Should we not view the population as varying about the "normal" level, X_s, rather than X_B? Mathematically, either will do, as long as they are close to one another in the appropriate sense. However, there is something to be gained by using X_s instead of X_B: The resulting equations are easier to handle. This can best be illustrated if we reformulate the above problem appropriately. But we shall leave this as an exercise for the reader and illustrate the technique in the more complicated situation of Section 3.7.

Exercises for Section 3.6

1. A certain species exists independently with an equilibrium population density $X_B = 1$. The parameter m of Eq. (4) which determines the rate of growth of population near the level X_B is $m = 1.1$. At year zero, a second species starts to prey on this one and takes a fixed population of $N = 0.05$ annually.
   (a) Use Eq. (13) to calculate the numbers $\xi_0, \xi_1, \xi_2, \xi_3$, and ξ_4 giving estimates of the prey population for the next four years.
   (b) Use Exercise (1a) of Section 3.5 to obtain a bound for the error $|\xi_n - x_n|$ valid for $n = 0, 1, 2, 3$, and 4.
   (c) Use Eq. (14) to calculate ξ_{10}. Compare this value with that of the new equilibrium population X_s determined by Eq. (15).

2. In the notation of this section, a species for which $X_B = 1.05$ and $m = 1$ is subject to the constant level of predation $N = 0.05$.
   (a) Use Eq. (16) to derive a quadratic equation for the new equilibrium population level X_s. Solve this to find X_s precisely.
   (b) Now calculate the *estimate* (15) of X_s. (The number X_s and its estimate should not differ by more than 0.0024.)

3. A process is described by the relation

$$4p_n = 3p_{n-1}^2 - 2p_{n-1} + 3 \qquad (n = 1, 2, 3, \ldots),$$

and p_0 is given. Find the constant (equilibrium) solutions for this relation (i.e., values of p, if any, for which $p_n = p$, $n = 0, 1, 2, \ldots$). Find a *linear* relation which yields a good approximation to solutions which are close to any constant solutions you may find. (*Hint:* Having found p, let $p_n = p + x_n$, where x_n is supposed to be small.)

4. In the numerical examples of Fig. 3.5, what would you *guess*, without calculation, that the behavior of the prey population would be if the predators disappeared after year 4?

5. Suppose that, in the analysis of Eq. (16), we take X_s as defined in Eq. (18) and define $y_n = X_n - X_s$ for each n. Prove that, if $\beta = 1 + mX_B - 2mX_s$, then for $n = 1, 2, \ldots$,

$$y_n = \beta y_{n-1} - my_{n-1}^2.$$

An approximating sequence for the y's is now obtained from

$$\eta_n = \beta\eta_{n-1},$$

which should be compared with Eq. (13).

3.7 Mutually Dependent Predator-Prey Species

One of our fundamental assumptions in the last model was the independence of the predator population from that of the prey. In this section, we set up a model for the situation in which the prey is the principal source of food for the predator. Thus, the population of predators will be directly controlled by the size of the population of prey. We now need a new variable, say Y_n, for the total population of the predator species in year n.

We now make our assumptions defining the relationship between the population levels. We might add that these are speculative in nature, and it would be surprising if the kind of model we are now considering were to predict anything more than the qualitative behavior of the population levels. We assume that, on the average, *the number of prey taken by any one predator is proportional to the population level of the prey*. Thus, for small variations about the equilibrium population of the prey, the diet of the predator improves or deteriorates with the availability of his principal source of food. The second assumption is that *the rate of growth of the predator species is proportional to the consumption of food per individual*. Thus, we suppose that more or less offspring are produced, depending on whether the predator is well fed or poorly fed.

Let S_n be the rate of growth of the predator species in year n. We interpret the two italicized statements above as saying that, since the food consumed is proportional to X_n, S_n is also proportional to X_n. Since X_B is fixed in any particular problem, we may write

$$S_n = \frac{k}{X_B}X_n$$

for each n, where k is some fixed positive number, and we have written the constant of proportionality as k/X_B for our subsequent convenience. Since $Y_n = S_{n-1}Y_{n-1}$, we now obtain

(19) $$Y_n = \frac{k}{X_B}(X_{n-1}Y_{n-1}) \qquad \text{(for } n = 1, 2, 3, \ldots)$$

as the recurrence relation governing the predator population. We see immediately that, if there are no prey in the year $n - 1$, then $X_{n-1} = 0$ and, consequently, $Y_n = 0$, i.e., the predators are annihilated in the subsequent year because they have no source of food.

The population of prey *in the absence of predators* is again supposed to be governed by the relation (5) which may be written:

$$X_n = (1 - mx_{n-1})X_{n-1}$$
$$= (1 - m(X_{n-1} - X_B))X_{n-1}$$

or

$$X_n - X_{n-1} = mX_B X_{n-1} - mX_{n-1}^2.$$

We have seen, however, that the number of prey lost to the predators in this year is a multiple of $X_{n-1}Y_{n-1}$. Thus, for some positive constant c, the governing equation for the prey may be written:

$$(20) \qquad X_n - X_{n-1} = mX_B X_{n-1} - mX_{n-1}^2 - cX_{n-1}Y_{n-1}.$$

The model is now complete. If we assume that the biologist can supply numerical values of c, m, k, and X_B, then given values of X_0, Y_0, we can put $n = 1$ in Eq. (19) and (20) and immediately calculate X_1 and Y_1. We then put $n = 2$ in these equations and use the calculated X_1, Y_1 to obtain X_2, Y_2, etc.

As a numerical example, consider a situation in which $m = 0.75$, $X_B = 1.5$, $k = 1.5$, and $c = 0.1$. The initial population densities are given by $X_0 = 1$ and $Y_0 = 4$. The variations in the populations predicted by Eqs. (19) and (20) for the subsequent ten years are illustrated in Fig. 3.6.

Before analyzing the problem mathematically, we first follow the suggestion made at the end of the previous section and ask if the two species could exist in a state of mutual equilibrium. That is, could the populations X_n, Y_n remain constant for a certain number of years (in the absence of other disturbing factors)? Suppose that X_E, Y_E are such equilibrium levels. Obviously $X_E = Y_E = 0$ gives an equilibrium case, but we assume $X_E \neq 0$ and $Y_E \neq 0$. If the populations in year $n - 1$ are at this level, they will be unchanged in year n, and we have

$$X_{n-1} = X_n = X_E, \qquad Y_{n-1} = Y_n = Y_E.$$

Substituting these equilibrium values in Eqs. (19) and (20), we obtain

$$(21) \qquad X_E = \frac{X_B}{k}, \qquad Y_E = \frac{m(k-1)}{ck}X_B.$$

If this is to make sense physically, we must have $Y_E > 0$ and hence $k > 1$. To see the physical significance of this, we re-examine Eq. (19). How

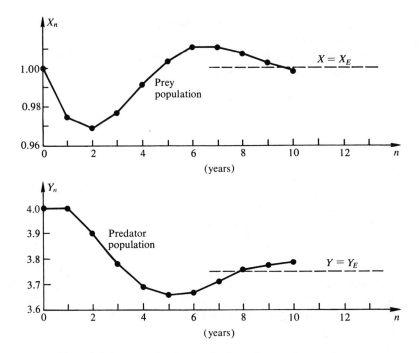

Figure 3.6 An example of mutually dependent predator and prey.

would we expect the predator population to behave if the prey population
was at the level X_B, the equilibrium population in the absence of predators,
and if there were only very few predators? In this case we put $X_{n-1} = X_B$
in Eq. (19) and are left with $Y_n = kY_{n-1}$; thus k is the rate of reproduction
of the predator in a situation where the food supply is plentiful. If, as we
assume, the reproduction rate depends only on food supply, then in this
condition we must expect a high rate of reproduction for the predator—a
rate higher than that needed to just maintain the population. In other words,
we expect on these physical grounds that k is indeed greater than 1.

In the numerical example of Fig. 3.6, we find that $X_E = 1$ and $Y_E = 3.75$.
We see from Fig. 3.6 that X_n and Y_n do appear to be approaching these
numerical values as n increases. Note also that the starting values $X_0 = 1$,
$Y_0 = 4$ imply that the initial prey population was at the equilibrium level
while that of the predator was high. This explains the initial drop in the
population of prey. However, the population of prey recovers when the
reduction in food for the predator has had time to take effect, i.e., less food
implies a smaller number of predators, hence less predation, and a recovery
in the number of prey.

We now proceed by reformulating Eqs. (19) and (20) in order to express
them in terms of variations in population about the equilibrium levels

X_E, Y_E given in Eq. (21). To this end, we define for each n,

$$x_n = X_n - X_E \quad \text{and} \quad y_n = Y_n - Y_E$$

and substitute $X_n = X_E + x_n$, $Y_n = Y_E + y_n$, $X_{n-1} = X_E + x_{n-1}$, $Y_{n-1} = Y_E + y_{n-1}$ in Eqs. (19) and (20). When this is done, the results simplify because of our definition of X_E and Y_E. The resulting equations can then be written in the form:

$$(22) \qquad x_n = (1 - mX_E)x_{n-1} - cX_Ey_{n-1} - mx_{n-1}^2 - cx_{n-1}y_{n-1}$$

$$(23) \qquad y_n = \left(\frac{Y_E}{X_E}\right)x_{n-1} + y_{n-1} + \frac{1}{X_E}x_{n-1}y_{n-1}.$$

Writing the equations in this way, we can see how to obtain a simplified pair of equations which will produce solutions showing the qualitative behavior of the solutions of Eqs. (22) and (23), provided the x's and y's are sufficiently small. We consider new equations obtained by omitting the small terms in Eqs. (22) and (23) which arise from products of x's and y's among themselves. Thus,

$$(24) \qquad \xi_n = (1 - mX_E)\xi_{n-1} - cX_E\eta_{n-1}$$

$$(25) \qquad \eta_n = \left(\frac{Y_E}{X_E}\right)\xi_{n-1} + \eta_{n-1}$$

$$(n = 1, 2, 3, \ldots)$$

and we take $\xi_0 = x_0$, $\eta_0 = y_0$.

It is obviously easier to compute with these equations than with Eqs. (19) and (20) but, what is more important, it is easier to make qualitative statements about the behavior of the numbers ξ_n and η_n as n increases from zero. These are determined by the coefficients of the simultaneous recurrence relations whose coefficients are represented by the array, or matrix,

$$\begin{bmatrix} 1 - mX_E & -cX_E \\ Y_E/X_E & 1 \end{bmatrix}.$$

We could use this example to motivate a study of matrix-vector algebra and obtain properties of the solutions by this means. However, we shall leave this to a later occasion and proceed in a different way.

Returning to Eqs. (24) and (25), we shall try to obtain a single recurrence relation involving only η's. To do this, we shall eliminate the ξ's from Eq. (24) by using Eq. (25). Recall that these equations are true for all $n \geq 1$, and so we deduce from Eq. (25) that

$$\xi_{n-1} = \frac{X_E}{Y_E}(\eta_n - \eta_{n-1}),$$

and

$$\xi_n = \frac{X_E}{Y_E}(\eta_{n+1} - \eta_n).$$

Substituting these two expressions in Eq. (24) and rearranging the resulting relation yields

(26) $\eta_{n+1} - (2 - mX_E)\eta_n + (1 - mX_E + cY_E)\eta_{n-1} = 0.$

We again have a single recurrence relation, but this time it is of quite a different type. In order to estimate the increment in the predator population (to calculate η_{n+1}) in year $n + 1$, we must know the corresponding values in the *two* preceding years. Thus, to start off the calculations, we need to know η_0 *and* η_1. However, given η_0 and ξ_0, this poses no problem because Eq. (25) yields

$$\eta_1 = \left(\frac{Y_E}{X_E}\right)\xi_0 + \eta_0.$$

Note also that the behavior of the η sequence, as predicted by Eq. (26), depends on only two numerical constants, $2 - mX_E$ and $1 - mX_E + cY_E$.

For the example of Fig. 3.6 it is found that $\eta_1 = \eta_0 = 0.25$, and for $n \geq 1$,

$$\eta_{n+1} - (1.25)\eta_n + (0.625)\eta_{n-1} = 0.$$

The values of η predicted by this relation, up to $n = 10$, give results so close

TABLE   3.1

Numerical Values for a Predator-Prey Model.

n	$X_n = x_n + X_E$	$\xi_n + X_E$	$Y_n = y_n + Y_E$	$\eta_n + Y_E$
0	1.000	1.000	4.000	4.000
1	0.975	0.975	4.000	4.000
2	0.969	0.969	3.900	3.906
3	0.977	0.977	3.779	3.789
4	0.991	0.990	3.692	3.701
5	1.003	1.002	3.659	3.665
6	1.010	1.009	3.671	3.674
7	1.010	1.010	3.708	3.708
8	1.007	1.007	3.747	3.745
9	1.002	1.002	3.772	3.770
10	.998	.999	3.779	3.778
15	1.001	1.001	3.741	3.741

to those of Fig. 3.6 that they cannot be distinguished on that sketch. Note that $y_n = Y_n - Y_E = Y_n - (3.75)$ and η_n is an approximation for y_n. Thus, $Y_n \simeq 3.75 + \eta_n$. The numerical values obtained are given in Table 3.1. (Note that the entries are rounded to three decimal places *after* being computed with higher accuracy.)

Exercises for Section 3.7

1. Verify that Eqs. (19) and (20) imply Eqs. (22) and (23).
2. Suppose that the population levels of mutually dependent predator-prey species are governed by

$$X_n - X_{n-1} = \tfrac{1}{2}X_{n-1} - \tfrac{1}{2}X_{n-1}^2 - \tfrac{1}{8}X_{n-1}Y_{n-1}$$
$$Y_n = 2X_{n-1}Y_{n-1} - Y_{n-1}^2.$$

   What are the equilibrium population levels for the two species?
3. Summarize the hypotheses of the mathematical model used in Section 3.7.
4. The recruiting officer at a certain institution is requested to bring the total number of staff up to N members. Appointments which are to be effective as of January 1st each year must be made by July of the preceding year, and he makes enough appointments to fill the vacancies the institution had at that time. Unfortunately, resignations are not handed in until December, and 10% of the employees resign each year. When the officer took over the job on January 1, 1970, the number of employees was $\tfrac{4}{5}N$. On January 1, 1971, he was quite pleased with his policy and felt that his method would soon bring the staff up to strength. On January 1, 1972, he was not so sure. Explain why this happened by calculating the number of employees on these dates, and show that his policy will eventually lead to a virtually constant staff which is understrength. What should he have done to bring the institution up to strength in 1971 and to keep it there?

3.8 Volterra's Principle

Let us take advantage of the predator-prey model constructed in the last section to examine a phenomenon recognized some time ago by Volterra. Suppose that the two species are coexisting at the equilibrium population levels X_E and Y_E given by Eq. (21). We imagine the possibility of some natural catastrophe which applies to *both* species and removes a large proportion of both populations without destroying either species entirely. The catastrophe might take several forms, e.g., forest fire, flood, incursion of a third species preying on both the others, or the overuse of insecticides harmful to both species. Will the populations recover subsequently? If so, in what way?

Volterra's principle suggests that, after the catastrophe, the population of the prey species will rapidly increase to exceed X_E, while the population of predators will decrease even further in subsequent years and the species may even be annihilated. The reasons for this short-term response are intuitively clear. When the predator population is very far below Y_E, the prey are able to multiply while being relatively unharmed by the predator. On the other hand, the initial reduction in prey means the removal of the predator's food supply and presents the danger of dying of starvation, even if one is lucky enough to escape the first catastrophe!

In the longer run, however, there is a possibility of survival for the predator. If the prey population increases fast enough, the food supply for the predator may reach a level of abundance which will save the predator and, further, allow them to increase. We might then expect a cyclic phenomenon to be exhibited (as in Fig. 3.6) and the two populations to return ultimately to their equilibrium levels.

We illustrate these ideas with the model and numerical example of the last section. Recall that the populations are controlled by the following values of the parameters: $m = 0.75$, $X_B = 1.5$, $k = 1.5$, $c = 0.1$. These give rise to equilibrium population levels (Eq. (21)) $X_E = 1$ and $Y_E = 3.75$. Let us suppose that the catastrophe removes a half of each population so that in year zero the populations are given by

$$X_0 = 0.5 \quad \text{and} \quad Y_0 = 1.875$$

or

$$x_0 = X_0 - X_E = -0.5 \quad \text{and} \quad y_0 = Y_0 - Y_E = -1.875.$$

Is the model appropriate? The hypothesis of the model which we are most likely to contravene is the assumption of a linear variation of the rate of growth of the prey population with the population level. This was discussed in Section 3.3, and as it was claimed there, this is plausible as long as the population level does not vary too much from its equilibrium value. We will proceed on the rather more drastic assumption that this variation of R with X is valid for a wider range of values of X than we have previously admitted. In particular, Eq. (3) (and hence Eq. (4)) is supposed to hold for all X between $\frac{1}{2}X_E$ and $\frac{3}{2}X_E$.

If we accept all the other hypotheses relating to the model and the foregoing numerical data, then we can use Eqs. (22) and (23) which now take the form:

$$x_n = (0.25)x_{n-1} - (0.1)y_{n-1} - (0.75)x_{n-1}^2 - (0.1)x_{n-1}y_{n-1}$$
$$y_n = (3.75)x_{n-1} + y_{n-1} + x_{n-1}y_{n-1}.$$

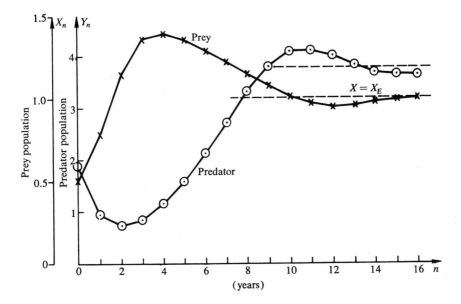

Figure 3.7 Population levels after a catastrophe.

Starting the computation with $x_0 = -0.5$ and $y_0 = -1.875$, we obtain the population levels illustrated in Fig. 3.7.

The phenomena described above are clearly illustrated. The prey population rapidly increases and exceeds the equilibrium level by almost 40% only four years after the catastrophe. However, by this time the predator population has recovered to such an extent that it subsequently cuts the prey down to size once more.

Immediately after the catastrophe, the predator species is in a bad way and reaches a minimum two years after the event. But by this time the prey are plentiful again, and so the predator species is able to recover. After a period of about fifteen years, both species have returned very close to their equilibrium population levels.

Exercises for Section 3.8

1. Why is it that Eqs. (24) and (25) (which are easier to work with) cannot be used in place of Eqs. (22) and (23) in the example of this section?
2. (a) Show that, for the numerical example of this section, Eq. (25) predicts a zero predator population one year after the catastrophe.
   (b) Suppose, for the sake of argument, that this really happens. How do you expect the population to behave subsequently? Is this what Eqs. (22) and (23) (or (24) and (25)) would predict?

3.9   Other Forms of Biological Association

We have now examined two models of pairs of species associated through the predator-prey relationship. That is, the first species feeds from the second, and the second is clearly disadvantaged by the presence of the first. Put in this way, it is intuitively clear that a similar mathematical formalism can be expected in the examination of host-parasite interactions.

Other forms of biological association of species can be modeled in similar ways. Let us briefly discuss a model of *competition* between two species for the same source of food. We suppose that both species depend entirely on the same food supply and that, if left in isolation, either species has a population level governed by the linearized model of Section 3.3. Thus, if X_n, Y_n are the population levels of the two species in year n ($n = 0, 1, 2, \ldots$), then without competition

$$(27) \qquad X_n = R_{n-1} X_{n-1} \quad \text{and} \quad Y_n = S_{n-1} Y_{n-1}$$

where, as in Eqs. (4) and (5),

$$(28) \qquad R_{n-1} = 1 - m(X_{n-1} - X_B) \quad \text{and} \quad S_{n-1} = 1 - \mu(Y_{n-1} - Y_B),$$

X_B, Y_B being stable equilibrium populations and m, μ positive constants.

We suppose that the loss of food supply from one species to the other is reflected in the birth rate. Thus, we need to reformulate Eqs. (28) to account for this as follows. Assume there exist positive constants, k and l, such that

$$(29) \qquad \begin{aligned} R_{n-1} &= 1 - m(X_{n-1} - X_B) - kY_{n-1} \\ S_{n-1} &= 1 - \mu(Y_{n-1} - Y_B) - lX_{n-1}. \end{aligned}$$

Then the larger the population of the first species (X_{n-1}), the less food is available for the second, and its rate of growth (S_{n-1}) is affected directly by a reduction of magnitude lX_{n-1}. A symmetrical situation is supposed to exist with respect to the other species.

The simultaneous recurrence relations governing the two population levels are now obtained by substitution from Eq. (29) into Eq. (27). The next step is to decide whether equilibrium population levels X_E, Y_E exist at which the two species could coexist. We do this as described in Section 3.7 for the mutually dependent predator-prey species. It is found that, if

$$k < \frac{mX_B}{Y_B} \qquad \text{and} \qquad l < \frac{\mu Y_B}{X_B}$$

then $m\mu > kl$ and there exist positive equilibrium populations

$$(30) \qquad X_E = \frac{mX_B - kY_B}{m\mu - kl}\mu, \qquad Y_E = \frac{\mu Y_B - lX_B}{m\mu - kl}m.$$

We now rewrite Eq. (27) in terms of increments x_n, y_n above X_E, Y_E, respectively. Thus

$$x_n = X_n - X_E \quad \text{and} \quad y_n = Y_n - Y_E \qquad (n = 0, 1, 2, \ldots)$$

and, after simplication

(31)
$$
\begin{aligned}
x_n &= (1 - mX_E)x_{n-1} - (kX_E)y_{n-1} - mx_{n-1}^2 - kx_{n-1}y_{n-1} \\
y_n &= -(lY_E)x_{n-1} + (1 - \mu Y_E)y_{n-1} - lx_{n-1}y_{n-1} - \mu y_{n-1}^2.
\end{aligned}
$$

For small variations in population levels about X_E, Y_E we could define approximations to the x_n any y_n by setting $\xi_0 = x_0$, $\eta_0 = y_0$ and then for $n = 1, 2, 3, \ldots$,

(32)
$$
\begin{aligned}
\xi_n &= (1 - mX_E)\xi_{n-1} - (kX_E)\eta_{n-1}, \\
\eta_n &= -(lY_E)\xi_{n-1} + (1 - \mu Y_E)\eta_{n-1}.
\end{aligned}
$$

The general features of the association between species can often be inferred by an examination of the coefficients appearing in the right-hand side of such recurrence relations. Suppose that for any such relation we write

(33)
$$
\begin{aligned}
\xi_n &= a_{11}\xi_{n-1} + a_{12}\eta_{n-1} \\
\eta_n &= a_{21}\xi_{n-1} + a_{22}\eta_{n-1}
\end{aligned}
\qquad (n = 1, 2, \ldots).
$$

In the case of competing species as represented by Eqs. (31), we have the information that $a_{12} < 0$ and $a_{21} < 0$. This is obtained immediately from Eqs. (31) since we assume k, l, X_E, and Y_E to be positive.

In contrast to this, the "coupling terms" in the predator-prey equations (24) and (25) are of opposite sign: $a_{12} < 0$ and $a_{21} > 0$. The fact that both a_{12} and a_{21} are negative in the case of species in competition simply reflects the fact that the presence of either species is detrimental to the other. In the predator-prey case, $a_{12} < 0$ implies that increases in predator population give rise to decreases in that of the prey. On the other hand, $a_{21} > 0$ implies that increases in prey population are beneficial to that of the predator.

Exercises for Section 3.9

1. Complete the detailed derivation of Eqs. (30) and (31).
2. The model of competition may, with some modifications, be applied to two different segments of a human population. How would a_{12}, a_{21} in Eq. (33) behave if we approach the extreme case of exploitation of one group by the other?
3. In the biological association known as *symbiosis*, the second species lives on the first but has a beneficial effect on it. If a recurrence of the form of Eq. (31) can be derived for these species, what signs do you expect a_{12} and a_{21} to have?

4. Set up and examine a model to describe the symbiotic relationship.

5. The population of a host species X and a parasite Y are related by the equations:

$$X_{n+1} = 1.12X_n - 0.1X_nY_n$$
$$Y_{n+1} = 0.95Y_n + 0.05X_nY_n.$$

Find the equilibrium levels, X_E and Y_E, of the two populations. Make the substitutions $X_n = x_n + X_E$, $Y_n = y_n + Y_E$ and obtain equations for the deviations x_n, y_n of the populations about their equilibrium levels.

3.10   Three-Term Recurrence Relations

In this section we take a brief look at some recurrence relations from the point of view of pure mathematics. Thus, we leave aside any physical motivation for studying recurrence relations and try to discover general properties of solutions for problems posed in the abstract. The relation we study depends on two fixed real numbers b and c and has the form

(34) $x_{n+2} + bx_{n+1} + cx_n = 0$ (for $n = 0, 1, 2, \ldots$).

We define a *solution* to be an infinite sequence of real numbers $x_0, x_1, x_2, \ldots$ such that Eq. (34) is true for any three consecutive numbers of the sequence.

Having reached this point in formulating the problem mathematically, we must ask whether our definition makes sense. That is, do any solutions exist? We will show that solutions do exist for all possible choices of b and c by exhibiting particular sequences and verifying the definition. This can be described as showing the existence by a *constructive* method, rather than seeking a direct "yes" or "no" answer to the existence question. However, in this case, the answer is clearly "yes" because we may specify x_0 and x_1 as *any* two real numbers whatever and then use Eq. (34) to compute $x_2, x_3, \ldots$ sequentially. In particular, it is clear that the sequence $0, 0, 0, \ldots$ is a solution. This is known as the *trivial* solution.

Consider two solutions, $y_0, y_1, y_2, \ldots$ and $z_0, z_1, z_2, \ldots$. This means that, for each integer $n \geq 0$,

$$y_{n+2} + by_{n+1} + cy_n = 0$$

and

$$z_{n+2} + bz_{n+1} + cz_n = 0.$$

Adding these two equations and defining a new sequence by $x_n = y_n + z_n$, we find that

$$x_{n+2} + bx_{n+1} + cx_n = 0   (n = 0, 1, 2, \ldots).$$

Thus, with the proper definition of the addition of sequences, we see that the sum of two solutions is again a solution.

Now we define the product of the sequence $y_0, y_1, y_2, \ldots$ with any real number α to be the sequence $\alpha y_0, \alpha y_1, \alpha y_2, \ldots$. If $y_0, y_1, y_2, \ldots$ is a solution of Eq. (34), then by multiplying both sides of the equation

$$y_{n+2} + by_{n+1} + cy_n = 0$$

by α, we easily see that the sequence defined by $x_n = \alpha y_n$ is once more a solution.

These two properties are characteristic of *linear* problems. We can combine the two properties in the following way: If $y_0, y_1, y_2, \ldots$ and $z_0, z_1, z_2, \ldots$ are solutions, then by the second property, so are $\alpha y_0, \alpha y_1, \alpha y_2, \ldots$ and $\beta z_0, \beta z_1, \beta z_2, \ldots$. Therefore, by the first property, their sum is a solution. The sum is now a sequence of the form

(35) $$x_n = \alpha y_n + \beta z_n \qquad (n = 0, 1, 2, \ldots).$$

where α, β may be any two real numbers.

Note that in Section 3.4 we approximated the solutions of a nonlinear problem (Eq. (6)) with those of a linear one (Eq. (7)). Similarly in Section 3.7, the problem as first posed (Eq. (22) and (23)) is nonlinear, and the approximating model (Eq. (24) and (25)) gives rise to a linear problem.

It is difficult to arrive directly at explicit nontrivial solutions of Eq. (34) without some prior knowledge of problems of this type. In order to use only the rather limited mathematical tools at our disposal, we have to assume considerable prior knowledge. In particular, we first look for solutions in the form of a geometric sequence with the first term 1 and common ratio λ:

(36) $$y_n = \lambda^n \qquad (n = 0, 1, 2, \ldots),$$

where λ is a nonzero real number. We don't do this because it is a natural approach, but only because the author's foreknowledge tells us that it leads to the most elementary mathematical treatment. If a solution has the given form, then the recurrence relation yields

$$\lambda^{n+2} + b\lambda^{n+1} + c\lambda^n = 0$$

and since $\lambda \neq 0$,

(37) $$\lambda^2 + b\lambda + c = 0.$$

Thus, if λ is a *real* root of this quadratic equation, then the sequence $y_0, y_1, y_2, \ldots$ is a solution. But it is only for certain combinations of b and c

that the quadratic has real roots. Thus we have a partial solution to the problem:

(1)   If $b^2 > 4c$ and

$$\lambda_1 = -\frac{b}{2} + \sqrt{\frac{b^2}{4} - c}, \qquad \lambda_2 = -\frac{b}{2} - \sqrt{\frac{b^2}{4} - c},$$

then we have two solutions given by

$$y_n = \lambda_1^n, \qquad z_n = \lambda_2^n \qquad (n = 0, 1, 2, \ldots),$$

and using Eq. (35) we may say that, for *any* real α and β, the sequence

(38) $$x_n = \alpha \lambda_1^n + \beta \lambda_2^n$$

is a solution of Eq. (34).

(2)   If $b^2 = 4c$, our quadratic equation has only one root, $\lambda_1 = -b/2$, and so we have one solution of the form $y_n = \lambda_1^n$. We can easily verify that another solution is given by $z_n = n\lambda_1^n$. We have

$$z_{n+2} + bz_{n+1} + cz_n = (n + 2)\lambda_1^{n+2} + b(n + 1)\lambda_1^{n+1} + cn\lambda_1^n$$
$$= \lambda_1^n\{(\lambda_1^2 + b\lambda_1 + c) + 2\lambda_1^2 + b\lambda_1\}.$$

Since $\lambda_1 = -b/2$, then $2\lambda_1^2 + b\lambda_1 = 0$ and, by definition of λ_1, $\lambda_1^2 + b\lambda_1 + c = 0$. Thus, $z_{n+2} + bz_{n+1} + cz_n = 0$. By using Eq. (35), we can again characterize a wide class of solutions, valid in the case $b^2 = 4c$, by

(39) $$x_n = (\alpha n + \beta)\lambda_1^n \qquad (n = 0, 1, 2, \ldots)$$

for any choice of real α and β.

(3)   If $b^2 < 4c$, then the roots of Eq. (37) are complex numbers, and it seems that the resulting sequences obtained from Eq. (36) will be complex and hence are not solutions of Eq. (34). (We defined solutions to be *real* sequences.) There is a way around this dilemma which allows us to use the complex roots to build up real sequences, but this requires a little more familiarity with complex numbers than we are prepared to assume. We therefore make do with verification that certain sequences are solutions of Eq. (34) when $b^2 < 4c$.

In this case we obviously have

$$c > 0 \quad \text{and} \quad \left|\frac{b}{2\sqrt{c}}\right| < 1.$$

So we can define a real angle θ between 0 and π by

$$\cos \theta = -\frac{b}{2\sqrt{c}}, \qquad \sin \theta = \sqrt{1 - \frac{b^2}{4c}}.$$

We also define $A = \sqrt{c} > 0$ and claim that two solutions of Eq. (34) are:

(40) $\qquad y_n = A^n \cos n\theta, \qquad z_n = A^n \sin n\theta \qquad (n = 0, 1, 2, \ldots).$

The verification of these statements involves the use of some trigonometric identities. We shall verify only the first solution and leave the second as an exercise for the reader. We need the following relations:

$$\cos 2\phi = \cos^2 \phi - \sin^2 \phi$$
$$\sin 2\phi = 2 \sin \phi \cos \phi$$
$$\cos (\phi + \psi) = \cos \phi \cos \psi - \sin \phi \sin \psi,$$

and note that they are true for *all* angles ϕ and ψ. We observe first of all that, for our particular angle θ,

$$\cos 2\theta = \cos^2 \theta - \sin^2 \theta = \frac{b^2}{4c} - \left(1 - \frac{b^2}{4c}\right) = \frac{b^2}{2c} = 1$$

$$\sin 2\theta = 2 \sin \theta \cos \theta = -\frac{b}{\sqrt{c}} \sqrt{1 - \frac{b^2}{4c}}$$

Now we have

$$y_{n+2} + by_{n+1} + cy_n = A^n \{A^2 \cos (n+2)\theta + bA \cos (n+1)\theta + c \cos n\theta\}$$
$$= A^n \{A^2 \cos n\theta \cos 2\theta - A^2 \sin n\theta \sin 2\theta$$
$$+ bA \cos n\theta \cos \theta - bA \sin n\theta \sin \theta + c \cos n\theta\}.$$

Substituting for $A = \sqrt{c}$ and for $\cos \theta$, $\sin \theta$, $\cos 2\theta$, and $\sin 2\theta$:

$$y_{n+2} + by_{n+1} + cy_n = A^n \left\{ c\left(\frac{b^2}{2c} - 1\right) \cos n\theta + c\frac{b}{\sqrt{c}} \sqrt{1 - \frac{b^2}{4c}} \sin n\theta \right.$$
$$+ b\sqrt{c} \left(-\frac{b}{2\sqrt{c}}\right) \cos n\theta$$
$$\left. - b\sqrt{c} \sqrt{1 - \frac{b^2}{4c}} \sin n\theta + c \cos n\theta \right\}.$$

Gathering the terms in $\cos n\theta$ and in $\sin n\theta$ on the right, we find that the right-hand side reduces to zero. Thus

$$y_{n+2} + by_{n+1} + cy_n = 0,$$

which is what we wanted to prove.

Taking for granted the solution $z_0, z_1, z_2, \ldots$ of Eqs. (40), we have solutions of the form

(41) $x_n = A^n(\alpha \cos n\theta + \beta \sin n\theta)$ $(n = 0, 1, 2, \ldots),$

in the case $b^2 < 4c$, for any real α and β.

We have now exhausted all possibilities of choices for b and c and can give solutions in the form of Eq. (38), (39), or (41), depending on the relative magnitudes of b^2 and $4c$. Another question which should now be asked is: Can there be any other solutions not covered by those classes of solutions we have found? Again, we are not in a position to answer this question since we don't have the necessary mathematical techniques. However, it is possible to prove that we have included in Eq. (38), (39), or (41) *all* solutions of Eq. (34).

Let us illustrate the usefulness of this analysis by formulating the explicit solution of the numerical example at the end of Section 3.7:

$$\eta_{n+2} - (\tfrac{5}{4})\eta_{n+1} + (\tfrac{5}{8})\eta_n = 0 \qquad (n = 0, 1, 2, \ldots),$$

with $\eta_0 = \eta_1 = \tfrac{1}{4}$. First we observe that, in the notation of this section, $b = -\tfrac{5}{4}, c = \tfrac{5}{8}$, and hence $b^2 < 4c$. Thus, the solution *must* have the form of Eq. (41) for some choice of α and β. To determine these numbers, we use the data $\eta_0 = \eta_1 = \tfrac{1}{4}$. We have

$$\eta_n = A^n(\alpha \cos n\theta + \beta \sin n\theta)$$

where

$$\cos \theta = \tfrac{1}{2} \cdot \tfrac{5}{4} \cdot \sqrt{\tfrac{8}{5}} = \sqrt{\tfrac{5}{8}}, \qquad \sin \theta = \sqrt{1 - \tfrac{1}{4} \cdot \tfrac{25}{16} \cdot \tfrac{8}{5}} = \sqrt{\tfrac{3}{8}}$$

and $A = \sqrt{c} = \sqrt{\tfrac{5}{8}}$. Hence

$$\eta_0 = \tfrac{1}{4} = \alpha$$
$$\eta_1 = \tfrac{1}{4} = \sqrt{\tfrac{5}{8}}(\sqrt{\tfrac{5}{8}}\alpha + \sqrt{\tfrac{3}{8}}\beta).$$

From these equations we find that $\alpha = \tfrac{1}{4}$ and $\beta = \tfrac{1}{4}\sqrt{\tfrac{3}{5}}$. Thus the complete solution is

$$\eta_n = \tfrac{1}{4}(\tfrac{5}{8})^{n/2}(\cos n\theta + \sqrt{\tfrac{3}{5}} \sin n\theta),$$

where, from trigonometric tables, $\tan \theta = \sqrt{\frac{3}{5}} \simeq 0.7746$, whence $\theta \simeq$ 37.76°. We could now use this formula to obtain the circled points in Fig. 3.6 and the figures in the last column of Table 3.1. Of course, we don't have to calculate them successively if this is not needed. (Recall that, in Fig. 3.6, $\eta_n \simeq Y_n - Y_E$.)

We remark, finally, that the qualitative nature of solutions of the recurrence relation (34) can be inferred from our general solutions. For example, if $b^2 > 4c$ and $b < 0$, then solution (38) obtains, and it is easily seen that λ_1 and λ_2 are both positive and $\lambda_1 > \lambda_2$. The solution then indicates that, for large enough n (if $\alpha \neq 0$), the part of the solution dependent on λ_1 will dominate and will represent a steadily increasing or decreasing sequence, depending on whether $\lambda_1 > 1$ or $\lambda_1 < 1$.

Solutions of the form in Eq. (41) obviously have an oscillating component resulting from the trigonometric factor. The magnitude of these oscillations is governed then by the factor $A^n = c^{n/2}$. Thus, the oscillations increase or decrease in magnitude depending on whether $c > 1$ or $c < 1$.

Exercises for Section 3.10

1. Find two distinct non-trivial solutions of

(a) $\qquad x_{n+2} - 3x_{n+1} + 2x_n = 0 \qquad (n = 0, 1, 2, \ldots),$

(b) $\qquad 2x_{n+2} + x_{n+1} - x_n = 0 \qquad (n = 0, 1, 2, \ldots),$

(c) $\qquad x_{n+2} - 4x_{n+1} + 4x_n = 0 \qquad (n = 0, 1, 2, \ldots).$

Then find the solution of each relation for which $x_0 = 1$ and $x_1 = 0$.

2. Prove that the sequence $z_0, z_1, z_2, \ldots$ of Eq. (40) is a solution of Eq. (34) when $b^2 < 4c$.

3. If

$$x_n = ax_{n-1} + by_{n-1}$$
$$y_n = cx_{n-1} + dy_{n-1} \qquad (n = 1, 2, \ldots),$$

prove that

$$x_{n+1} - (a + d)x_n + (ad - bc)x_{n-1} = 0 \qquad (\text{for } n = 1, 2, 3, \ldots).$$

If $a = 4$, $d = 1$, $b = 2$, and $c = 0$, show (by looking for solutions of the form $x_n = \lambda^n$) that, for any real α and β,

$$x_n = \alpha + \beta 4^n \qquad (n = 0, 1, 2, \ldots)$$

determines a solution.

4. Construct perpendicular axes in the plane using coordinates b and c. Indicate the regions of the plane corresponding to points b and c for which $b^2 > 4c$, $b^2 = 4c$, and $b^2 < 4c$. Prove your results.

5. Show that the recurrence relations (33) imply

$$\eta_{n+2} - (a_{11} + a_{22})\eta_{n+1} + (a_{11}a_{22} - a_{12}a_{21})\eta_n = 0 \quad \text{(for } n = 0, 1, 2, 3, \ldots).$$

Compare this with the relation (34) and prove that, if a_{12} and a_{21} are either both negative (*competing* species) or both positive (*symbiosis*) then $b^2 > 4c$. (In these cases, all solutions of the above difference equation are of the form

$$\eta_n = \alpha\lambda_1^n + \beta\lambda_2^n \quad (n = 0, 1, 2, \ldots)$$

for real $\alpha, \beta, \lambda_1, \lambda_2$; see Eq. (38).)

6. Garbage accumulates outside a factory at a rate of 5 tons per day. Three garbage collection teams are employed, each coming every three days, no two on the same day, and always early in the morning. No team wants to work harder than the others, so only one third of the garbage is removed each day. Set up a difference equation for the amount of garbage left in the evening, and show that it tends to a constant value which you should find.

7. (Use the notation and terminology of Section 3.4.) Suppose that the rate of increase of a certain species in year n depends not only on the population level in year n but also on the level in year $n - 1$. Thus, a linearized rate of increase for year n has the form

$$R_n = 1 - m_1 x_n - m_2 x_{n-1}.$$

Obtain the linear relation for an approximating problem:

$$\xi_{n+2} - (1 - m_1 X_B)\xi_{n+1} + m_2 X_B \xi_n = 0.$$

If $m_1 = 2$, $m_2 = \frac{1}{2}$, and $X_B = 1$, show that the solution of this problem has the form

$$\xi_n = (\tfrac{1}{2})^{n/2}(\alpha \cos n\theta + \beta \sin n\theta) \quad (n = 0, 1, 2, \ldots)$$

where $\cos \theta = -1/\sqrt{2}$, $\sin \theta = 1/\sqrt{2}$. (You may quote results of Section 3.10.) Then show that if $\xi_0 = 0$, $\xi_1 = \frac{1}{10}$, then

$$\xi_n = \tfrac{1}{5}(\tfrac{1}{2})^{n/2} \sin n\theta,$$

and calculate $\xi_0, \xi_1, \xi_2, \ldots, \xi_6$.

3.11   Concluding Remarks

In closing this chapter, it is probably wise to recall some of the shortcomings of our models for population growth, and only then to suggest some possible lines of further development. First, we have confined our attention to *discrete* models in that we consider the size of populations at a sequence of discrete points in time (for convenience, we have used yearly intervals of time in the body of the chapter). For some species, this is not unreasonable. For other species which are numerous and do not have a pronounced seasonal growth pattern (humans, for example), it may be more appropriate to view population and rates of growth as functions of time. Thus, they can be sampled or evaluated at *any* instant of time. In such a case, we obtain models involving rates of change of population levels in the resulting equations. The study of these models requires knowledge of the calculus rather than recurrence relations. Since we have not assumed knowledge of calculus and do not wish to present it here, we have deliberately avoided such models. They do, however, exhibit phenomena very closely related to those we have observed in this chapter.

Another question we have glossed over concerns the measures used for population levels. In some cases, the population may consist of creatures in quite distinct stages of development; for example, some species of flies go through the egg, larva, pupa, and adult stages. In other cases it is argued that a count of reproducing females is more significant than a total count. At the other end of the spectrum, "biomass" may be the most appropriate measure.

Another factor which we have not admitted and which may be relevant in long-term studies concerns the possibility of evolution by adaption. For example, a species which is constantly preyed upon by a predator, but is not exterminated, may ultimately develop a defense mechanism which will alleviate the predation. This can come about by the classical evolutionary process of adaption by natural selection. A similar intriguing phenomenon is the possibility of social action in a species (human or ant, for example) in the face of a common danger. The danger could arise from predation or competition with other species. It is questionable whether such effects can be included in mathematical models in a realistic way.

In our discussions of the effect of population density on the rate of growth of populations, we have assumed that the availability of food, for example, is an important controlling mechanism. This may be assumed to typify *external* controls on the population density. There is, however, strong experimental evidence that, in some cases, this is *not* a controlling mechanism. Overcrowded animals with an abundant food supply may yet have a low birthrate. This leads to the intriguing idea that species that have survived may have done so because they have evolved an innate *internal* control mechanism which tends to prevent large deviations in the population density from some optimum level. For example,* humans under stress (from overcrowding, perhaps) reproduce at less than the normal rate, and their ability to cure damaged tissue is impaired. This idea is particularly interesting because,

*See, for example, J. A. Gray, *The Psychology of Fear and Stress*, Weidenfeld and Nicolson, London, 1971.

in the short term, these effects would tend to remove from the population those individuals in whom they were most apparent. How, then, could these very traits appear in a highly evolved species through the process of natural selection?

Finally, we mention two possibilities for mathematical studies beyond those included so far. First, it may be necessary to examine the interactions of more than two species in some instances. Furthermore, it may be necessary to include the growth of plant species in certain models. For example, in a study of coyote and rabbit populations including the effect of a large fire of grassland, it may be necessary to include the quantity of grass, coyotes, and rabbits as principal variables. This kind of approach leads to the study of "systems ecology." Second, we note that all the problems studied have been of an "initial-value" type, that is, given data for the problem at an initial instant of time, we trace the future development. Other important uses of these models may include the possibility of *control*. Thus, if certain variables are under the control of the experimenter, the theoretician may be able to predict how the controllable variables should be assigned in order to arrive at certain desirable population levels after a fixed period of time. This kind of problem is typical of an area of applied mathematics known as "control theory." The combination of control theory with systems ecology offers some exciting prospects for combinations of theoretical and experimental studies.

Appendix I

The approximations used on page 54 and in other chapters can be made more precise with the following theorem. This is a special case of an important result from analysis known as Taylor's theorem. It tells us something about the error which is made in approximating the binomial expression $(1 + t)^\alpha$ where $|t| \leq \frac{1}{2}$ by the much simpler expression $1 + \alpha t$.

Theorem. *If t is a real number with* $-\frac{1}{2} \leq t \leq \frac{1}{2}$ *then for any real* α,

$$|(1 + t)^\alpha - (1 + \alpha t)| \leq \frac{|\alpha(\alpha - 1)|}{2} \frac{t^2}{(1 + t)^{2-\alpha}}$$

Note that, because of the t^2 factor on the right, the approximation of $(1 + t)^\alpha$ by $1 + \alpha t$ for a fixed α improves considerably as $|t|$ approaches zero.

We shall not attempt a proof, but for help in understanding the result, the reader should check the following:

1. Show that we can replace $\leq$ by $=$ in the theorem in the cases:

$$\alpha = 0, 1, 2 \text{ and all } t,$$
$$t = 0 \qquad \text{and all } \alpha.$$

2. If $0 \leq t \leq \frac{1}{2}$ and $\alpha \leq 2$ then the right-hand side can be replaced by the simpler bound $\frac{1}{2}|\alpha(\alpha - 1)|t^2$ because $1/(1 + t)^{2-\alpha} \leq 1$.

3. Show that $1/\sqrt{1.1}$ differs from 0.95 by less than 0.00375. Check this by calculating $1/\sqrt{1.1}$ to four decimal places. (Hint: Write $1/\sqrt{1.1} = (1 + 0.1)^{-1/2}$ and apply the theorem with $t = 0.1$ and $\alpha = -\frac{1}{2}$. Use item 2. above.)

4. Show that 0.99 raised to the power $\frac{1}{100}$ differs from 0.9999 by less than 0.0000006.

5. Verify the approximation used on page 54.

Preliminary Exercises

1. Express as integers:

$$1^6, \quad 4^4, \quad 10^3, \quad (\tfrac{1}{2})^{-2}, \quad (\sqrt{3})^4, \quad (2^3)(\tfrac{1}{2})^2, \quad ((\tfrac{1}{3})^{-2})^2.$$

Express as rational numbers:

$$4^{-2}, \quad (\tfrac{1}{2})^3, \quad (\tfrac{4}{9})^{1/2}, \quad (3 \times \tfrac{1}{2})^3, \quad (10^2)^{-3}.$$

Express in decimal form:

$$(2)^{-4}, \quad (0.1)^2, \quad (0.1)^6, \quad (0.1)^{10}, \quad ((0.1)^5)^2, \quad (\tfrac{1}{3})^{-5}(\tfrac{1}{3})^4.$$

(Recall the laws of exponents: If x is any nonzero real number and m, n are integers, then $x^0 = 1$ and

$$x^m \cdot x^n = x^{m+n}, \quad (x^m)^n = x^{mn}, \quad (xy)^n = x^n y^n.)$$

2. (a) (Recall that each real number corresponds to a point on the real line, and conversely.) What are the distances between the following pairs of points on the real line:

$$(0, 2), \quad (0, -2), \quad (1, 2), \quad (-1, 2), \quad (4, -\tfrac{3}{2}), \quad (-1, -\tfrac{1}{2})?$$

(b) Express as nonnegative real numbers:

$$|2 - 0|, \quad |-2 - 0|, \quad |2 - 1|, \quad |-2 - (-1)|,$$
$$|-\tfrac{3}{2} - 4|, \quad |-\tfrac{1}{2} - (-1)|.$$

3. Draw sketches of the solution sets of the following conditions; the universe is the real numbers:

$$|x| = 1, \quad |x| \leq 5, \quad |x - 2| \leq 1, \quad |x - 2| \leq \tfrac{1}{10},$$
$$|x + \tfrac{1}{2}| \leq 2, \quad |2x + 2| \leq 4.$$

4. (a) A hundred dollars is put into a savings account and accumulates interest at 4% per year (compounded annually). How much is in the account after 1, 2, 3, 4 years? (If p is the sum deposited at a rate of r percent per year for n years, then the amount is given by $p(1 + r)^n$.)

(b) If inflationary effects imply a reduction in the real value of savings of 5% per year, what is the real value of the savings after 1, 2, 3, 4 years?

5. Let S_n be the sum of the first n terms of a geometric series (progression) with first term a and common ratio r. Thus,

$$S_n = a + ar + ar^2 + \cdots + ar^{n-1}.$$

Multiply both sides of this equation by r and subtract the resulting equation from the first. Show that, if $r \neq 1$, then

$$S_n = a\left(\frac{1 - r^n}{1 - r}\right).$$

6. Use the result of the preceding question to find

$$2 + 1 + \tfrac{1}{2} + \tfrac{1}{4} + \tfrac{1}{8}, \qquad 3 + 6 + 12 + 24 + 48 + 96,$$

$$9 - 3 + 1 - \tfrac{1}{3} + \tfrac{1}{9} - \tfrac{1}{27} + \tfrac{1}{81}$$

and

$$1 + \tfrac{1}{10} + (\tfrac{1}{10})^2 + \cdots + (\tfrac{1}{10})^n$$

for $n = 5, 10$, and 20. What are the last three sums evaluated to four decimal places of accuracy?

7. The radiator of an automobile contains six gallons of water. Two gallons are drawn off and replaced by alcohol. Two gallons of the resulting mixture are then drawn off and replaced by alcohol. This is repeated several times. Establish the formula giving the proportion a_n of alcohol in the radiator after n repetitions:

$$a_n = \tfrac{1}{2}(1 - (\tfrac{1}{3})^n).$$

8. Sketch graphs of the functions f, g, and h defined on the real numbers by

$$f(x) = x^2 - 1, \qquad g(x) = 3x^2 + 1, \qquad h(x) = -4x^2 + 16$$

9. Using the functions defined in Preliminary Exercise 8 find (algebraically) those real values of x for which $f(x) = 0$, $g(x) = 0$, $h(x) = 0$. In other words, find the solution sets in the real numbers of each of the conditions $f(x) = 0$, $g(x) = 0$, $h(x) = 0$. In still other words, solve the quadratic equations $f(x) = 0$, $g(x) = 0$, $h(x) = 0$. Confirm your results by reference to the graphs sketched in Preliminary Exercise 8.

10. Solve the following quadratic equations by factoring:

$$x^2 + x - 2 = 0, \qquad 2x^2 - x - 3 = 0$$

$$4x^2 + 4x - 3 = 0, \qquad 12x^2 + 27x + 6 = 0.$$

11. Since $(2x + 1)^2 = 4x^2 + 4x + 1$, we may solve one of the above equations by completing the square. First observe

$$4x^2 + 4x - 3 = 0 \Rightarrow 4x^2 + 4x + 1 = 3 + 1.$$

Now finish the argument to find the real solutions of $4x^2 + 4x - 3 = 0$.

12. Every real quadratic polynomial has the form $ax^2 + bx + c$ for some real a, b, c with $a \neq 0$. Show that

$$ax^2 + bx + c = 0 \Rightarrow x^2 + \frac{b}{a}x + \frac{b^2}{4a^2} = \frac{b^2}{4a^2} - \frac{c}{a}.$$

Hence show that the real solutions of $ax^2 + bx + c = 0$ (if any) are given by

$$x = \frac{-b \pm \sqrt{b^2 - 4ac}}{2a}.$$

13. Use the last formula to confirm the results of Preliminary Exercises 9 and 10.

14. Let J be the set of all positive integers and $P_1, P_2, \ldots$ be statements which may be true or false. Thus for *each* $n \in J$ (and there are infinitely many of them), P_n is such a statement. We can try to prove that all of the statements are true by the method of *proof by induction* as follows:

Step 1: Prove P_1 true.
Step 2: Make the *induction hypothesis* that, for any fixed positive integer $v, P_1, P_2, \ldots, P_v$ are true.
Step 3: Use step 2 to prove that P_{v+1} is true.
Conclusion: P_n is true for every $n \in J$.

In practice, we often compute the first few statements, say P_1, P_2, P_3, P_4. Then we make a *guess* at what P_n should be and attempt to prove the conjectured P_n true by induction. If this does not work out, we *may* see in the course of step 3 how the conjectured P_n should be modified.

Let the statement P_n represent Eq. (1) of Section 1.5. Verify that P_n is true for $n = 1, 2, 3, 4, 5$, and prove by induction that it is true for all $n \in J$.

15. Prove by induction that, for every integer n, $n^5 - n$ is divisible by 5. (*Hint:* Verify the truth of the statement when $n = 1$. Make the *induction hypothesis* that the statement is true for integers $1, 2, \ldots, v$. Then consider the statement for $v + 1$ and use the binomial theorem to show that P_v true implies P_{v+1} true.)

16. Prove by induction that, for all integers $m \geq 1$,

$$\frac{1}{(1)(3)} + \frac{1}{(3)(5)} + \cdots + \frac{1}{(2m - 1)(2m + 1)} = \frac{m}{2m + 1}.$$

CHAPTER FOUR

SEARCHING FOR
A MAXIMUM

4.1 Introduction

How does a manufacturer of Ping-Pong balls decide how many balls he should manufacture each year? Suppose that his sole objective is to make as big a profit as possible. The larger the number he produces, the lower the production cost per ball is likely to be. On the other hand, these costs will have to be balanced by sales, and he knows that, if he produces too many balls, he will not sell them all and there will be waste. Thus, viewed as a function of volume of production only (although many other factors will be involved), the manufacturer must look for that volume of production which will *maximize* his profit. This plausible line of argument suggests that profit as a function of production yields a graph which is concave down, as illustrated in Fig. 4.1. The manufacturer must try to estimate P_0.

There is nothing special about Ping-Pong balls, of course, and the question can be posed for any manufacturing process.

We shall investigate procedures for estimating P_0 which are, at first sight, unsophisticated. The idea is that the manufacturer is allowed a certain limited number of trials with differing levels of production; he can find the profit corresponding to each trial and he must use this information to estimate P_0. We shall call each trial calculation of profit for a given production an *experiment*.

In real life there are two ways in which he might arrive at his experiments. They may be historical and simply consist of checking the manufacturer's records in comparable preceding years. He may also perform an experiment by using a mathematical model for his business and, making various assump-

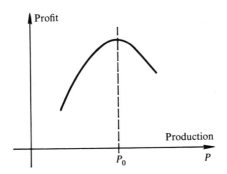

Figure 4.1 Profit and production.

tions regarding other factors, estimate the profit by a (possibly involved) calculation starting from a certain hypothesis on the level of production.

The question is this: If it is known that only a limited number of experiments can be made, how should they be chosen (at what production levels) in order to give the best estimate of P_0 (that level of production which will maximize profit)?

There are a great many situations in which problems of this kind arise. The *law of diminishing returns* in economics provides appropriate examples. A classical form of this runs as follows, although it appears in many other situations. It is supposed that a certain fixed area of land is available for food production, and we wish to determine the amount of labor that should be allocated for the purpose. We are interested in the amount of production (measured in bushels of wheat, say) as a function of the size of the labor force employed.

First, there is obviously no production if there is no labor. Suppose we measure labor in "units of labor," which may mean one man working one hour, or a hundred men working one week, in different problems. By utilizing one unit of labor, it is found that a certain production is obtained. What will be the effect of adding a second unit of labor? Will the production be doubled? The law of diminishing returns says that production will *not* be doubled; the production per unit of labor will be somewhat less than that for one unit of labor. Furthermore, this effect is said to multiply as more units of labor are added. Indeed, as suggested by Fig. 4.2, there may well be a point at which total production begins to fall as too many units of labor are added. It is clear that, all other factors remaining unchanged, the fixed area of land implies that there will be a certain maximum possible production which cannot be exceeded whatever labor force is used. Thus, a linear growth of production with labor would not be possible. In this case, we are to estimate the amount of labor needed to maximize production.

In Chapter 3 we met a function which was also concave down (Fig. 3.2).

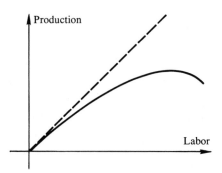

Figure 4.2 Law of diminishing returns.

We can use this to suggest a more specific example that may be of interest to a farmer: What population density of a poultry farm will result in the highest rate of reproduction?

What are the essential mathematical features of these problems? In each case we are given a *function* defined on a real interval, and in this interval the function has a *unique maximum* value. The function is an *increasing* function of x to the left of the maximum and a *decreasing* function of x to the right of the maximum. We now formalize these ideas in a definition:

A function y of the variable x is *unimodal on the interval* $[a, b]$ if:

1. There is a unique x^* in $[a, b]$ such that

$$\max_{a \leq x \leq b} y(x) = y(x^*).$$

2. If x_1, x_2 are in $[a, b]$ and $x_1 < x_2$, then

$$x_2 < x^* \quad \text{implies} \quad y(x_1) < y(x_2)$$
$$x^* < x_1 \quad \text{implies} \quad y(x_1) > y(x_2).$$

Part 1 of the definition tells us that the function y does indeed have a maximum in $[a, b]$, and part 2 tells us that y is an increasing function of x to the left of the maximum and a decreasing function to the right. Hence there is only one point in $[a, b]$ at which y attains its maximum value $y(x^*)$. We shall write $y^* = y(x^*)$.

Readers who have studied differential calculus will have met another technique for finding maxima of certain unimodal functions. This classical method is confined to functions which are *differentiable* on some interval. In our search techniques, we assume only that the function can be calculated for any x in $[a, b]$. The function may not have a derivative at some points of the interval as indicated in Fig. 4.3(c). In Fig. 4.3(a), we have tried to

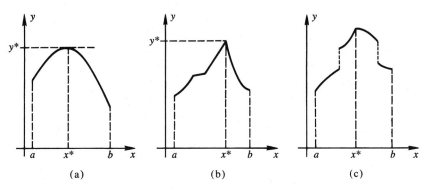

Figure 4.3 Unimodal functions.

indicate the graph of a function which is differentiable; in (b) the function is continuous but not differentiable, and in (c) the function is not continuous but is still unimodal.

Before proceeding to analyze our model, we can introduce a simplifying idea. We may assume, without loss of generality, that $[a, b] = [0, 1]$. If, for any particular problem we have $a \neq 0$ or $b \neq 1$, then we can make a simple change of variable so that this is achieved. Thus (with $a < b$) we define a new variable ξ by

$$\xi = \frac{x - a}{b - a}$$

so that when $x = a$, $\xi = 0$, and when $x = b$, we have $\xi = 1$. We then make the substitution

$$x = (b - a)\xi + a$$

in order to obtain y as a function of the independent variable ξ. This process corresponds geometrically to first shifting the position of the origin along the x axis (see Fig. 4.3) and then changing the scale of the variable on the horizontal axis. We do not wish to labor this point, so let us accept the result that after such a transformation, a function y which was a unimodal function of x on $[a, b]$ will be a unimodal function of ξ on $[0, 1]$. This can be proved directly using the definition of a unimodal function. Henceforth, we only consider functions which are unimodal on $[0, 1]$.

We now have our model: a unimodal function f on $[0, 1]$. The problem is to say, as best we can, where the maximum of the function lies using only finitely many "experiments", that is, using only finitely many function evaluations in $(0, 1)$. We have seen that there is a variety of real-life situations for which the model is appropriate.

When we consider the arrangement of our experiments, there are two features which arise immediately. First is the spatial distribution of the experiments in [0, 1]. For example, if we are permitted four experiments, we must choose the values x_1, x_2, x_3, x_4 in [0, 1] at which to evaluate the function f. The second feature is the arrangement of these experiments in time. We first select x_1 and then, taking account of the calculated value of $f(x_1)$, we select x_2 and calculate $f(x_2)$; then using knowledge of $f(x_1)$, $f(x_2)$, we select x_3; and so on. Here we are following a *sequential search* pattern. In some real-life situations, this may not be possible and, before we calculate any function values, the values x_1, x_2, x_3, and x_4 must be selected *simultaneously*. This is called a *simultaneous search*.

With a fixed number of experiments, a sequential search will generally be more efficient than any simultaneous search. However, both are of interest and of practical significance. In particular, examination of simultaneous search techniques will yield generally applicable strategies, independent of the function f.

Exercises for Section 4.1

1. The following sketch indicates revenue and cost as functions of level of production for a particular manufacturing process. Revenue is based on a fixed sale price per item produced and so increases linearly with the volume of production for low production levels. However, there is a certain saturation level of production beyond which no further sales, and hence revenue, can be expected. The costs increase with the production, of course, but the cost per item decreases initially as production increases. Sketch the graph of profit (revenue minus cost) versus production, and observe that it is unimodal on the interval from the break-even production level to the saturation level.

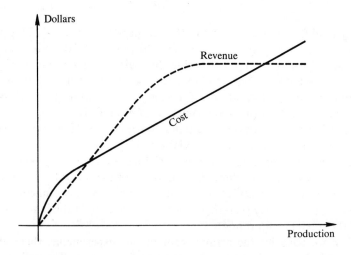

Figure 4.4

2. Which of the following functions f are unimodal on $[0, 1]$? Sketch a graph in
each case.
(a) $f(x) = x(1 - x)$
(b) $f(x) = x^2$
(c) $f(x) = \begin{cases} 2x & \text{(for } 0 \le x \le \frac{1}{2}) \\ 1 - x & \text{(for } \frac{1}{2} < x \le 1) \end{cases}$
(d) $f(x) = \begin{cases} 2x & \text{(for } 0 \le x \le \frac{1}{2}) \\ 0 & \text{(for } \frac{1}{2} < x \le 1) \end{cases}$

3. (This is an exercise in formal proof.) Let f be an increasing function of x on
$[a, b]$ and write $y = f(x)$. Let g be the linear function of ξ defined by
$$g(\xi) = (b - a)\xi + a.$$
Show that the function of ξ defined by $y = f(g(\xi))$ is an increasing function of
ξ on $[0, 1]$.

4.2 Simultaneous Search Techniques

It is clear that, if only one experiment is permitted, then no information is
obtained regarding the location of the maximum. We therefore begin by
examining how we can best locate two experiments simultaneously. Suppose
they are to be at x_1 and x_2, $0 < x_1 < x_2 < 1$. In Fig. 4.5 we indicate the
possible outcomes with regard to the relative magnitudes of $y_1 = f(x_1)$
and $y_2 = f(x_2)$. What conclusions can be drawn in each of these cases with
regard to the location of the maximum of f? In each case we will find that
the maximum must lie in a smaller *interval of uncertainty* than the interval
$[0, 1]$ itself.

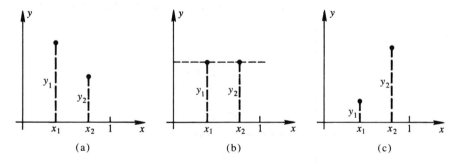

(a)   (b)   (c)

Figure 4.5 Possible outcomes of two experiments.

In case (a), we claim that the maximum y^* must be attained at a point
x^* in $[0, x_2]$, for if x^* were in $[x_2, 1]$, the function f could not be an increasing
function of x in $[0, x^*]$ as the unimodal property demands. Thus, with this
outcome of the experiments, the interval of uncertainty now has length
$l_a = x_2$.

In case (*b*), x^* must belong to (x_1, x_2), for if x^* is in $[0, x_1]$, the fact that $y_1 = y_2$ contradicts the fact that f is decreasing to the right of x^*, and if x^* is in $[x_2, 1]$, we contradict the increasing property of f to the left of x^*. In this case the interval of uncertainty has length $l_b = x_2 - x_1$.

Case (c) is similar to case (a) and yields an interval of uncertainty of length $l_c = 1 - x_1$.

Our problem is to locate x_1, x_2 in such a way that the new interval of uncertainty is as small as possible and the size of the interval will be valid for all unimodal functions f on $[0, 1]$. Thus, if we define

$$L_2 = \max (l_a, l_b, l_c) = \max (x_2, x_2 - x_1, 1 - x_1)$$

we wish to make L_2 as small as possible by choice of x_1, x_2 in $[0, 1]$. Since $x_2 - x_1 < x_2$, we can immediately write

$$L_2 = \max(x_2, 1 - x_1).$$

Now consider a fixed x_1 and view L_2 as a function of x_2. For a fixed x_1, when does $L_2(x_2)$ take its minimum value? We see (Fig. 4.6) that the least value, $1 - x_1$, is attained for any x_2 between zero and $1 - x_1$. In particular, for fixed x_1, the least value of $L_2(x_2)$ is attained when $x_2 = 1 - x_1$. Since we also have $x_1 < x_2$, this implies $x_1 < 1 - x_1$, i.e., $x_1 < \frac{1}{2}$.

Thus, for a fixed x_1, the least value of L_2 is attained when $\frac{1}{2} - x_1 = x_2 - \frac{1}{2} > 0$, and we then have

$$\min_{x_2} L_2(x_1, x_2) = 1 - x_1 > \frac{1}{2}$$

This means that x_1, x_2 should be symmetrically placed in $[0, 1]$ with respect

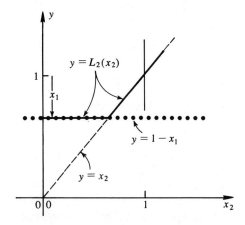

Figure 4.6 L_2 as a function of x_2.

to $x = \frac{1}{2}$. If we now try to reduce L_2 further by choice of x_1 (subject to $x_1 < \frac{1}{2}$), we find that x_1, x_2 should be chosen as close to $\frac{1}{2}$ as possible.

This argument also tells us that the length of the interval of uncertainty after two experiments must exceed $\frac{1}{2}$, but that it can be arbitrarily close to $\frac{1}{2}$! It cannot be exactly $\frac{1}{2}$ for we must have $x_1 < x_2$. Furthermore, $x_1 = x_2$ implies only one experiment and no information at all on the location of the maximum.

Our solution to the two-experiment problem is therefore to locate the pair symmetrically in the interval and as close as possible to the center.

How close is it possible to put them? The mathematician might be tempted to reply: "Closer than any real number $\epsilon > 0$ which you care to mention, however small ϵ may be." But this is of no practical value. The question can only be answered if we know more about the physical situation we are trying to model. In practice, we can generally specify a number $\epsilon > 0$ which is the least separation between experiments for which a significant difference in function values $y_1 = f(x_1)$ and $y_2 = f(x_2)$ is found. Once such an ϵ is specified, we locate the two experiments at

$$x_1 = \frac{1}{2} - \frac{\epsilon}{2} \quad \text{and} \quad x_2 = \frac{1}{2} + \frac{\epsilon}{2}$$

so that $x_2 - x_1 = \epsilon$, the least significant separation, and the interval of uncertainty is then $1/2 + \epsilon/2$. That is, whatever the outcome of the two experiments may be, we can find an interval of length $1/2 + \epsilon/2$ in which the maximum is known to lie. The quantity ϵ is sometimes known as the *resolution* of the process.

Another factor may come into play in determining ϵ if f can be computed only to finite accuracy—as is usually the case. Experiments must not be placed so close to one another that, to the accuracy of computation, corresponding values of f may coincide. Such a situation is illustrated in Exercise 3 of Section 4.3.

4.3 *Three Simultaneous Experiments*

We now consider the case of three simultaneous experiments and use the same technique as before. We can put the outcomes of any set of three experiments in one of the three cases illustrated in Fig. 4.7. In cases (a), (b), and (c) the first, second, and third experiments give the largest functional values, respectively. We can include the possible cases of pairs of equal functional values as indicated in the second column of the following table. Using the definition of a unimodal function, we now see very easily that the

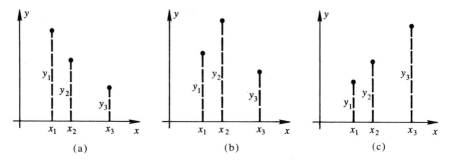

Figure 4.7 Possible outcomes of three experiments.

Case	Conditions	Length of Interval of Uncertainty
(a)	$y_1 \geq y_2 > y_3$	x_2
(b)	$y_1 < y_2,\ y_2 > y_3$	$x_3 - x_1$
(c)	$y_1 < y_2 \leq y_3$	$1 - x_2$

maximum of f in $[0, 1]$ must lie in the pair of subintervals on either side of the maximum experimental value obtained. This gives rise to the entries in the last column of the table. The worst result that can happen for any function f, then, is that the maximum of f lies in an interval of uncertainty of length

$$L_3 = \max(x_2, x_3 - x_1, 1 - x_2).$$

By choice of x_1, x_2, x_3, we must make L_3 as small as possible.

We first consider the possibility of reducing L_3 by choice of x_2. We note that if $x_2 \neq \frac{1}{2}$, then $\max(x_2, 1 - x_2) > \frac{1}{2}$, for we must have either x_2 or $1 - x_2$ greater than $\frac{1}{2}$. Furthermore, when $x_2 = \frac{1}{2}$, $\max(x_2, 1 - x_2) = \frac{1}{2}$. Thus, the least value which $\max(x_2, 1 - x_2)$ can have is $\frac{1}{2}$, and this is attained when $x_2 = \frac{1}{2}$. Thus,

$$L_3 \geq \max\{\tfrac{1}{2}, x_3 - x_1\}.$$

The best we can do now to keep L_3 as small as possible is to make sure that $x_3 - x_1 \leq \frac{1}{2}$.

The outcome of this argument is: Choose $x_2 = \frac{1}{2}$ and x_1, x_3 in *any* way, provided only that $x_3 - x_1 \leq \frac{1}{2}$ and $x_1 < x_2 < x_3$. We can then ensure that the length of the resulting interval of uncertainty does not exceed $\frac{1}{2}$.

In contrast to the case of two experiments, we can find a distribution of points for which this best possible result is actually attained. However, although we use one extra experiment, we are only able to get a marginal improvement in the length of the interval of uncertainty.

For example, if we describe a distribution as being "minimax", when the least value of L_3 is attained, then $x_1 = 0.4$, $x_2 = 0.5$, $x_3 = 0.7$ is minimax. However, since we generally prefer distributions with the greatest possible symmetry, we favor the minimax solution:

$$x_1 = \tfrac{1}{4}, \qquad x_2 = \tfrac{1}{2}, \qquad x_3 = \tfrac{3}{4}.$$

In the case of two points, there is no minimax solution unless we introduce the minimum ϵ-spacing. In such a case, the solution is called "ϵ-minimax".

Exercises for Section 4.3

1. Which pairs of points in [0, 1] are ϵ-minimax if (a) $\epsilon = 0.04$, and (b) $\epsilon = 0.1$?

2. Which of the following triples in [0, 1] are minimax:

$$(0.45, 0.5, 0.7), \qquad (0.2, 0.5, 0.78), \qquad (0.1, 0.5, 0.6)?$$

3. Consider the function f defined on [0, 1] by

$$f(x) = \tfrac{1}{10}x(1 - x)$$

and which is unimodal on [0, 1] (cf. Exercise 2 of Section 4.1). Suppose values of f are obtained by a computer which works only to three-decimal-place accuracy. What is the least significant spacing, ϵ, for experiments with f?
[*Hints for solution:* Pairs of points x, y for which $0 \leq x, y \leq \tfrac{1}{2}$ and $x - y \geq \epsilon$ must satisfy $f(x) - f(y) \geq 0.001$. These inequalities yield $x(1 - x) - y(1 - y) \geq 0.01$, whence

$$x - y \geq \frac{0.01}{1 - (x + y)}$$

and hence $\epsilon^2 \geq 0.01$ and $\epsilon \geq 0.1$.]

4.4 General Simultaneous Search Schemes

We now make a rather alarming leap from two and three experiments to any number of experiments. The essential idea which we employ, and which can easily be proved formally, is that the maximum of f must lie between the two experiments on either side of that experiment producing the largest value of f. For this purpose we include $x = 0$ and $x = 1$ as experimental values of x, although a function evaluation is not to be made at either of these points. If n experiments are made, we write $x_0 = 0$ and $x_{n+1} = 1$ and we deduce from the set of simultaneous experiments that the resulting interval of uncertainty is $[x_{k-1}, x_{k+1}]$ for some k between 1 and n. Thus, x^* will certainly lie in an interval of uncertainty of length

$$L_n = \max_{1 \leq k \leq n} (x_{k+1} - x_{k-1}),$$

and by choice of $x_1, x_2, \ldots, x_n$, we must make L_n as small as possible.

We consider first any even number, say $2n$, of experiments. We look for a search technique assuming that, as in the case $n = 1$, the points will be placed in pairs at the least possible distance apart, ϵ. This is called a search in *uniform pairs*.

We consider placing these pairs about the n points

$$x = \frac{1}{n+1}, \frac{2}{n+1}, \ldots, \frac{n}{n+1}.$$

Probably the most natural idea is then to put pairs of points at distance $\epsilon/2$ from each of the n points as indicated in (a) of Fig. 4.8. For this scheme we have

$$L_{2n} = \frac{1}{n+1} + \frac{\epsilon}{2}$$

(a)
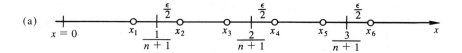

(b)
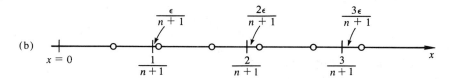

Figure 4.8 Search by uniform pairs.

and this biggest interval is attained at $k = 1$, i.e., $L_{2n} = x_2 - x_0 = x_2$. But this is obviously not a minimax arrangement because, by moving the first pair of points a little to the left, we can reduce L_{2n} a bit further. After a little thought, we then come up with scheme (b). Here the points are in ϵ-pairs, but they are distributed about the points $j/(n + 1), j = 1, 2, \ldots, n,$ as indicated in the diagram. In this case we see that

$$L_{2n} = \frac{1}{n+1} + \frac{\epsilon}{n+1} = \frac{1+\epsilon}{n+1}$$

and this is the width of *every* interval

$$[x_{k-1}, x_{k+1}] \qquad (\text{for } k = 1, 2, \ldots, 2n).$$

We claim that this distribution of points is ϵ-minimax. This can be verified by observing that no point, x_j, of the distribution can be moved (maintaining the minimum spacing) without *increasing* the value of L_{2n}.

Our experience with three experiments leads us to expect a rather different situation with an odd number of experiments. Generalizing from our limited experience we now suggest that, if there are $2n + 1$ experiments, the even-numbered experiments should be located at the evenly spaced points:

$$x_{2j} = \frac{j}{n+1} \qquad (j = 0, 1, 2, \ldots, n + 1).$$

With these points fixed, the points x_{2j+1} can be placed in any way consistent with the conditions that

$$x_{2j+1} - x_{2j-1} \leq \frac{1}{n+1} \qquad (j = 1, 2, \ldots, n + 1).$$

In particular, a uniform spacing of all $2n + 3$ points would satisfy this condition. This is illustrated in Fig. 4.9(a) in the case $n = 3$. However, many other asymmetric distributions are also possible and such a case is illustrated in (b) of the figure.

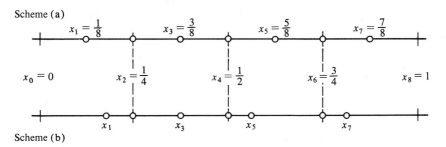

Scheme (a)

Scheme (b)

Figure 4.9 Searches with seven experiments.

Schemes of this type will always yield

$$L_{2n+1} = \frac{1}{n+1}.$$

We claim that all such schemes are minimax and observe that, with $2n + 1$ points, we have only a marginal improvement over the scheme with $2n$ points, as measured by the length of the resulting interval of uncertainty.

EXAMPLE. This example will illustrate several techniques numerically. Bear in mind that this example is artificial and is used here only for the purpose of illustrating the placing of experiments. The maximum of the unimodal function used can, in fact, be found more readily with the use of calculus. However, we have chosen the function so that the problem is not too trivial, even in that case.

Let

(1) $$f(x) = x\sqrt{\sin \pi x}$$

and suppose we are given that the function f is unimodal on [0, 1], as indeed it is.

Consider first a *simultaneous search with six experiments* and a minimum spacing between experiments of 0.02. The x values for a search in uniform pairs (see Fig. 4.8(b)) are in the first row of the following table. The point $x = \frac{1}{4}$ is between the first pair, $x = \frac{1}{2}$ is between the second pair, and $x = \frac{3}{4}$ between the third pair. The function values are then calculated to four decimal places, and we see that $f(x_5)$ is the

n	1	2	3	4	5	6
x_n	0.235	0.255	0.49	0.51	0.745	0.765
$f(x_n)$	0.1928	0.2161	0.4899	0.5099	0.6313	0.6276

largest value of f computed. Hence the interval of uncertainty is $[x_4, x_6] = [0.51, 0.765]$, which has length 0.255.

A search with *seven experiments* has already been discussed (Fig. 4.9) and, in this case, we may distribute the points evenly through the interval. The values of f for the function given by Eq. 1 are in the last row of the next table.

n	1	2	3	4	5	6	7
x_n	0.125	0.25	0.375	0.5	0.625	0.75	0.875
$f(x_n)$	0.077 3	0.210 2	0.360 4	0.5	0.600 7	0.630 7	0.541 3

In this case we have an interval of uncertainty $[x_5, x_7] = [0.625, 0.875]$ of length 0.25.

Exercises for Section 4.4

1. Let f be a unimodal function on [0, 1]. In the following table values of f are given at certain values of x on (0, 1). What is the interval of uncertainty?

x	0.1	0.23	0.4	0.51	0.8	0.95
$f(x)$	0.51	0.73	1.21	1.25	1.01	0.76

2. What is the ϵ-minimax location of four experiments in [0, 1] if $\epsilon = \frac{1}{8}$? What is the length of the resulting interval of uncertainty?

3. Which of the following locations of five experiments in [0, 1] are minimax:

$$\left(\tfrac{1}{6}, \tfrac{1}{3}, \tfrac{7}{12}, \tfrac{2}{3}, \tfrac{3}{4}\right), \qquad \left(\tfrac{1}{4}, \tfrac{1}{3}, \tfrac{1}{2}, \tfrac{2}{3}, \tfrac{3}{4}\right)?$$

4. Tabulate the values of x you would use to find the maximum of a unimodal function on [2, 3] in a simultaneous search with (a) eight experiments, and (b) nine experiments. The minimum distance between experiments is to be 0.02.

4.5 Dichotomous Search

We now turn to sequential search techniques. In this case it is assumed that in placing any experiment, we may take advantage of knowledge of the function values of all earlier experiments. Thus, in placing the experiment number n, say, we need only consider the interval of uncertainty obtained in the preceding $n - 1$ experiments.

Observe first of all that, if only two experiments are permitted, the result of the first gives us no information which helps in placing the second. Thus, in this case, there is no difference between simultaneous and sequential search, and the best we can do is the ϵ-minimax solution of Section 4.3 in which the resulting interval of uncertainty has width $L_2 = 1/2 + \epsilon/2$. Notice that the same principle applied to an initial interval of width, w would yield $L_2 = w/2 + \epsilon/2$.

The name "dichotomous" implies a continual dividing into two and that provides the clue to this search technique. We proceed by placing ϵ-pairs of experiments, rather than single experiments. We know that after two experiments we can locate the maximum in an interval of length $L_2 = 1/2 + \epsilon/2$. The advantage of a sequential search lies in the fact that we know *which* interval it is. In Fig. 4.10 we suppose that it turned out to be in the right-hand interval. We now place a second ϵ-pair in the middle of this interval, and the

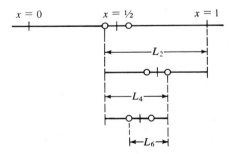

Figure 4.10 A dichotomous search.

resulting interval of uncertainty has width

$$L_4 = \frac{1}{2}\left(\frac{1}{2} + \frac{\epsilon}{2}\right) + \frac{\epsilon}{2} = \frac{1}{2^2} + \left(\frac{1}{2} + \frac{1}{2^2}\right)\epsilon.$$

In the sketch we suppose that this is at the left end of the second interval of uncertainty. We now put the third pair in the middle of this interval and obtain

$$L_6 = \frac{1}{2}(L_4) + \frac{\epsilon}{2} = \frac{1}{2^3} + \left(\frac{1}{2} + \frac{1}{2^2} + \frac{1}{2^3}\right)\epsilon.$$

Now it is very easily proved by induction that, after n pairs of experiments, we have

$$L_{2n} = \frac{1}{2^n} + \left(\frac{1}{2} + \frac{1}{2^2} + \cdots + \frac{1}{2^n}\right)\epsilon,$$

and summing the geometric series (see Preliminary Exercises 5 and 6 of Chapter 3),

(2) $$L_{2n} = \frac{1}{2^n} + \left(1 - \frac{1}{2^n}\right)\epsilon = \frac{1}{2^n}(1 - \epsilon) + \epsilon.$$

This is obviously very much better than our simultaneous technique for an even number of experiments. For example, after twenty experiments, if we disregard the ϵ-contributions, the intervals of uncertainty are $\frac{1}{11}$ and $\frac{1}{1024}$ in the simultaneous and dichotomous searches, respectively. It seems apparent from the above expression for L_{2n} that by choosing n large enough we can, in theory, make L_{2n} as close to the minimum spacing, ϵ, as we please. However, there is a limit on the number of experiments imposed by ϵ itself. This is explored in Exercises 2 and 3 below.

EXAMPLE.   Consider the problem posed at the end of the preceding section, i.e., to investigate the maximum of the function (1) with resolution $\epsilon = 0.02$ by dichotomous search.

The first pair of points are ϵ apart and evenly spaced in the middle of [0, 1]. Thus $x_1 = 0.49$ and $x_2 = 0.51$. On calculating $f(x_1)$ and $f(x_2)$, we find that $f(x_2)$

Pair Number	Experiment at	Function Values	Interval Uncertainty	Length
1	$x_1 = 0.49$ $x_2 = 0.51$	$f(x_1) = 0.4899$ $f(x_2) = 0.5099$	[0.49, 1.0]	0.51
2	$x_3 = 0.735$ $x_4 = 0.755$	$f(x_3) = 0.6321$ $f(x_4) = 0.6290$	[0.51, 0.755]	0.245
3	$x_5 = 0.6225$ $x_6 = 0.6425$	$f(x_5) = 0.5993$ $f(x_6) = 0.6100$	[0.6425, 0.755]	0.1125

$> f(x_1)$ and the resulting interval of uncertainty is recorded in the fourth column of the table. The length of the interval of uncertainty appears in the final column. The points x_3, x_4 are now placed in the middle of [0.49, 1.0], giving the coordinates tabulated. We now repeat the process and find that, after six experiments, we have an interval of uncertainty of length 0.1125.

Exercises for Section 4.5

1. A firm is attempting to maximize profit by varying the ratio of production between two articles. Since detailed accounting is costly, it can be certain of which scheme is better only when two test ratios differ by at least 0.05. Assuming that the profit is a unimodal function of the ratio, how many experiments are necessary for (a) a simultaneous search, and (b) a dichotomous search to obtain an interval of uncertainty of 0.12?

2. Prove that, with a minimum spacing of ϵ, the number of pairs of experiments in a dichotomous search on [0, 1] is limited by

$$n \leq \frac{\log (1 - \epsilon) - \log 4\epsilon}{\log 2}$$

   [*Hints for solution:* Deduce from a generalization of Fig. 4.10 that $L_{2n} \geq 5\epsilon$. Then use Eq. (2) to deduce $2^n \leq (1 - \epsilon)/4\epsilon$ and, hence, the result.]

3. Use the result of Exercise 2 to find the maximum number of pairs of experiments for the following values of ϵ: 0.1, 0.01, 0.001, 0.0001.

4. The most economical *equal interval* search technique is sometimes known as *trisection*. First compute three experiments evenly spaced in [0, 1]—i.e., at $\frac{1}{4}$, $\frac{1}{2}$, and $\frac{3}{4}$. Call the new interval of uncertainty I, and observe that it has width $\frac{1}{2}$. Trisect I, as above (note that one experiment is carried over from the preceding step), to find a new interval of uncertainty I_2. What is its width? If the process is repeated n times to obtain an interval of uncertainty I_n, what is its width?

5. How many trisection steps can be performed if the minimum spacing of experiments is ϵ?

6. Using dichotomous search, and then trisection, find the first three intervals of uncertainty for the maximum of the function

$$y = 3 - (x - 0.6)^2$$

   starting with the interval [0, 2].

4.6 Search by Golden Section

Consider now a general search technique in which j experiments have been completed, where $j \geq 2$. Where shall we place the $(j + 1)$th experiment? Using the principle enunciated at the beginning of Section 4.4, we can say that after j experiments the interval of uncertainty will have length $L_j = x_{k+1} - x_{k-1}$ for some k between 1 and j, and $f(x_k)$ is the largest (or one of the two largest) function values obtained in the j experiments. The position is therefore that indicated in Fig. 4.11. We want to know where x_{j+1} should be placed in this interval, which already contains three experiments: two at the end points and the maximum experimental value of f having been found at the interior point. One of these three is x_j, but we don't know (and don't need to know) which it is. For convenience, we suppose the left-hand end point is at $x = \xi_j$ and suppose the interior point is at a distance ξ from ξ_j. We shall also assume for convenience that $\xi < \frac{1}{2}$. It will be quite clear subsequently that a parallel argument can be developed if $\xi > \frac{1}{2}$, although we do wish to exclude the possibility $\xi = \frac{1}{2}$.

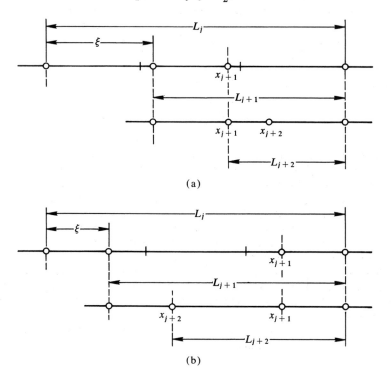

(a)

(b)

Figure 4.11 (a) The placing of x_{j+1}, x_{j+2} when $\frac{1}{3} < \xi < \frac{1}{2}$. (b) The placing of x_{j+1}, x_{j+2} when $0 < \xi < \frac{1}{3}$.

Suppose we place x_{j+1} at a distance d from the right-hand end of our interval of length L_j. Then we have

$$L_{j+1} = \max(L_j - \xi, L_j - d)$$

and we want to choose d to make L_{j+1} as small as possible. Since the length ξ is already fixed, we can do no better than choosing $d = \xi$. Thus, the suggestion is that *we choose x_{j+1} so that the two interior points are symmetrically placed in the interval.* Then we have $L_{j+1} = L_j - \xi$.

We can now apply the same principle to locate x_{j+2} in the interval of uncertainty of length L_{j+1}. Two possible positions are sketched in Fig. 4.11(a) and (b) on the assumption that x_{j+1} gave a maximum experimental value of the function. In Fig. 4.11(a), we suppose that $\frac{1}{3} < \xi < \frac{1}{2}$, and it is easily seen that, in the situation sketched, $L_{j+2} = L_j - (L_j - \xi)$ and also $L_{j+1} = L_j - \xi$. Combining these, and assuming the same "symmetric placing" rule is used throughout, we have

(3) $\qquad\qquad L_{j+2} + L_{j+1} - L_j = 0 \qquad$ (for $j = 2, 3, 4, \ldots$).

It is easily verified that this relation holds whatever the outcome of experiments x_{j+1} and x_{j+2} may be, and provided ξ is in the range $\frac{1}{2} < \xi < \frac{2}{3}$.

We shall suppose that the two initial steps of a search are located in such a way that this procedure can be used to determine x_3. From this we shall see that the whole sequence of experiments is determined and can be continued as long as accuracy permits.

Returning now to Fig. 4.11(b), we see that this situation can result if $0 < \xi < \frac{1}{3}$ (and a similar one if $\frac{2}{3} < \xi < 1$). In this case Eq. (3) does not hold. We shall leave this case as an exercise for the reader.

Equation (3) already tells us a lot about the method, regardless of the way in which it is started. This is a recurrence relation of the kind studied in Section 3.10 with $b = 1$ and $c = -1$. Thus $b^2 = 1 > -4 = 4c$ and the sequence of lengths is given by Equation (38) of Chapter 3:

(4) $\qquad\qquad L_j = \alpha \lambda_1^j + \beta \lambda_2^j \qquad (j = 2, 3, 4, \ldots)$

for some choice of α and β. The numbers λ_1, λ_2 are the solutions of $\lambda^2 + \lambda - 1 = 0$, namely

$$\lambda_1 = \frac{\sqrt{5} - 1}{2} \quad \text{and} \quad \lambda_2 = -\frac{\sqrt{5} + 1}{2}.$$

Observe that $1 > \lambda_1 > 0$ and $\lambda_2 < 0$ with $|\lambda_2| > 1$. If we have a solution (4) with $\beta \neq 0$, then the contributions of $\beta \lambda_2^j$ and $\beta \lambda_2^{j+1}$ will have opposite signs. Furthermore, the absolute value of this term increases with j, while that of $\alpha \lambda_1^j$ decreases with j. If we seek a method which will be valid for any number of experiments, then since we must always have $L_j > 0$, we must have $\beta = 0$ in Eq. (4) and we are left with solutions of the form

(5) $\qquad\qquad L_j = \alpha \lambda_1^j, \qquad \lambda_1 = \frac{\sqrt{5} - 1}{2}.$

With such a search procedure we have

(6)
$$\frac{L_{j+1}}{L_j} = \frac{1}{2}(\sqrt{5} - 1),$$

and so the ratio of the lengths of succeeding intervals of uncertainty is constant. We can see now that we should be able to improve on the dichotomous search since we have

$$\frac{L_{j+2}}{L_j} = \lambda_1^2 = \frac{1}{2}(3 - \sqrt{5})$$

and, since $2 < \sqrt{5} < 3$, $L_{j+2}/L_j < \frac{1}{2}$. In the dichotomous search, we got an improvement in the width of uncertainty of $\frac{1}{2}$, or more, at the expense of adding another pair of points.

Our process is still not complete, however, because we have not investigated the starting procedure. We have so far supposed $j \geq 2$. How should x_1, x_2 be located? We shall assume that they are to be symmetrically placed in [0, 1] with $\frac{1}{3} < x_1 < \frac{1}{2}$. Then Eq. (3) holds with $j = 1$; therefore (see Fig. 4.12)

$$L_3 + L_2 = 1$$

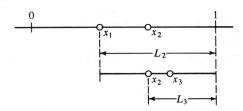

Figure 4.12 Placing the first experiments.

and also, by Eq. (6), $L_3 = \lambda_1 L_2$. Thus $(1 + \lambda_1)L_2 = 1$, or

$$L_2 = \frac{1}{1 + \lambda_1}.$$

Since $\lambda_1^2 + \lambda_1 - 1 = 0$ implies $\lambda_1(1 + \lambda_1) = 1$, we see that $1/(1 + \lambda_1) = \lambda_1$, and so we have $L_2 = \lambda_1$. Thus, we must have

$$x_1 = 1 - \lambda_1 = \tfrac{1}{2}(3 - \sqrt{5})$$
$$x_2 = \lambda_1 = \tfrac{1}{2}(\sqrt{5} - 1).$$

Now it only remains to determine α in (5) and our description of the search technique is complete. To do this we simply put $j = 2$ in Eq. (5) to obtain

$L_2 = \alpha \lambda_1^2$. But we have now chosen $L_2 = \lambda_1$ and so $\alpha \lambda_1^2 = \lambda_1$, whence $\alpha = 1/\lambda_1$ and we have the simple result

(7)
$$L_j = \lambda_1^{j-1}.$$

This is initially seen to be true for $j = 2, 3, \ldots$, but since we know that $L_1 = 1$, the result is true for $j = 1, 2, ,3, \ldots$.
Since $\sqrt{5} = 2.236\,067\,97\ldots$, the numerical value of λ_1 is

$$\lambda_1 = 0.618\,033\,98.\ldots$$

If the following table, we compare the intervals of uncertainty obtained after 4, 8, 12, 16, and 20 experiments in (a) a simultaneous search, (b) a dichotomous search, and (c) a search by golden section. We ignore the ϵ-contributions in the first two cases.

Number of Experiments	L_n, Width of Interval of Uncertainty		
n	Simultaneous	Dichotomous	Golden Section
4	$\frac{1}{3}$	$\frac{1}{4}$	$\frac{1}{4.236}$
8	$\frac{1}{5}$	$\frac{1}{16}$	$\frac{1}{29.03}$
12	$\frac{1}{7}$	$\frac{1}{64}$	$\frac{1}{199.0}$
16	$\frac{1}{9}$	$\frac{1}{256}$	$\frac{1}{1364.0}$
20	$\frac{1}{11}$	$\frac{1}{1012}$	$\frac{1}{9349.0}$

We see that after twenty experiments we can improve on the dichotomous search by a factor of almost $\frac{1}{10}$ on the width of the interval of uncertainty.

Why the name "golden section"? Suppose we take a segment of length 1 and divide it into segments of length l_1, l_2 with $l_2 > l_1$. How do we choose the point of division so that the ratio of lengths $1 : l_2$ is the same as the ratio $l_2 : l_1$? For us, this is a simple problem, although it was far from simple for the ancient Greeks who first showed an interest in it. We observe that

$$l_1 + l_2 = 1 \quad \text{and} \quad \frac{1}{l_2} = \frac{l_2}{l_1}.$$

From the second equation we have $l_1 = l_2^2$, and substituting in the first, we obtain the quadratic equation for l_2:

$$l_2^2 + l_2 - 1 = 0.$$

The only positive solution of this equation is $l_2 = \frac{1}{2}(\sqrt{5} - 1)$, just the number λ_1 of our search scheme. This ratio is thought to be particularly pleasing aesthetically and has been used in architectural design for centuries. For example, it has been used to determine the "ideal" ratio of height to width of window openings. The number has a variety of interesting properties, and in earlier times a certain aura of mysticism was attached to it. The word "golden" seems to be used, therefore, in the sense of "good", "valuable", and even "magical". The term "golden section" appears to have been in use for about 125 years.

EXAMPLE. We now apply search by golden section to the problem investigated at the conclusion of Sections 4.4 and 4.5. We shall use a decimal approximation to the "magic number" $\lambda_1 = \frac{1}{2}(\sqrt{5} - 1)$. We work to four decimal places and so we have $\lambda_1 \simeq 0.6180$. Consequently, choose $x_2 = 0.6180$ and, since x_1, x_2 are symmetrically placed in [0, 1], $x_1 = 0.3820$. Since $f(x_2) > f(x_1)$, we now place x_3 so that x_2, x_3 are symmetrically placed in [0.382, 1.0]. This yields $x_3 = 0.7640$. Now we find $f(x_3) > f(x_2)$, thus x_4 is fixed so that x_3, x_4 are symmetrically placed in [0.6180, 1.0]. This yields $f(x_4) < f(x_3)$, and so on. After six experiments, the interval of uncertainty has length 0.090 which is, of course, $(0.6180)^5$.

n	Experiment	Function	Interval of Uncertainty	Length
1	0.382	0.3688	[0, 1]	1
2	0.618	0.5967	[0.382, 1.0]	0.618
3	0.764	0.6278	[0.618, 1.0]	0.382
4	0.854	0.5683	[0.618, 0.854]	0.236
5	0.708	0.6309	[0.618, 0.764]	0.146
6	0.674	0.623	[0.674, 0.764]	0.090

Exercises for Section 4.6

1. Prove that, with a minimum spacing of ϵ, the number of experiments n in a search by golden section on [0, 1] is limited by

$$n \leq \frac{\log \epsilon - \log (\sqrt{5} - 2)}{\log \lambda_1} + 1$$

[Hints for solution: Deduce from a sketch such as in Fig. 4.11(a) that, after n experiments, the minimum distance between pairs of points is $2L_n - L_{n-1}$.

Then use (5) to obtain

$$2L_n - L_{n-1} = \lambda_1^{n-1}(\sqrt{5} - 2) \le \epsilon,$$

and hence the result.]

2. Use the result of the preceding exercise to find the maximum number of experiments for the following values of ϵ: 0.1, 0.01, 0.001, 0.0001. Compare the results with those of Exercise 3 of Section 4.5.

3. Show that with the distribution of points and maxima leading to the situation of Fig. 4.11(b), the lengths of successive intervals of uncertainty satisfy

$$L_{j+2} - 2L_{j+1} + L_j = 0.$$

If this is satisfied for $j = 1, 2, \ldots$, prove that, for some real α and β,

$$L_j = \alpha j + \beta.$$

[*Hint:* Use Eq. (38) of Chapter 3.] If $\alpha = 0$, then this solution implies that all L_j are equal, which is nonsense. So we must have $\alpha \ne 0$. But this implies that $|L_j|$ increases for large enough j. How is this explained?

4.7 An Example

We return to the manufacturer of Ping-Pong balls mentioned at the beginning of the chapter. We suppose that for any given production level, x, of balls per year, the accounting department can, with the aid of a computer program, provide a numerical value of profit $p(x)$.

It is known from experience that the maximum profit can be expected to be attained for x between half a million and one million and that p is a unimodal function on this interval. It is also known that there are such uncertainties built into the estimates used to write the program evaluating p, that a difference in production level of 25,000 balls per year is required before a significant difference in profit can be forecast. Thus, on a scale of 10^6, we have a resolution of 25,000.

From the modeling point of view, all the idealization and hypotheses are now contained in the "black box" of the accountants' program. The model will have no added validity as a result of computer programming, and the conclusions will have to be consistent with past history to be acceptable. Since we have not discussed the details of the model yielding profit as a function of production, we will not be able to go more deeply into this aspect of the problem.

For step 3, we choose to apply the search by golden section on the interval $[(\frac{1}{2})(10)^6, 10^6]$ with a minimum spacing of experiments of $\epsilon = (0.025)$

$(10)^6$. We are to predict as small a range of production levels as possible in which the optimum production level (i.e., that yielding maximum profit) is known to lie.

We begin by transforming the interval of interest onto the unit interval. Thus, as described in Section 4.1, we define a new variable ξ by

$$\xi = [x - (\tfrac{1}{2})(10)^6][10^6 - (\tfrac{1}{2})(10^6)]^{-1}$$
$$= (2x)(10)^{-6} - 1.$$

Then $x = (\tfrac{1}{2})(10)^6$ corresponds to $\xi = 0$ and $x = 10^6$ corresponds to $\xi = 1$.

We must also record the effect of this transformation on the minimum spacing of experiments. We observe here that an interval of length $(0.025)(10)^6$ on the x-scale transforms to an interval of length 0.05 on the ξ-scale. Thus, on the ξ-scale we take $\epsilon = 0.05$.

We begin the search by seeking values of p at ξ-values of 0.382 and 0.618 (i.e., at $1 - \lambda_1$ and λ_1 to four-decimal accuracy). The accountant will not recognize our ξ-variable, so we had better ask for the corresponding value of x obtained from $x = \tfrac{1}{2}(\xi + 1)10^6$. The first two production levels in question are, therefore, $(0.691)(10)^6$ and $(0.809)(10)^6$.

n	ξ	$10^{-6}x$	$10^{-5}p$	
1	0.382	0.691	0.75	
2	0.618	0.809	0.87	
3	0.764	0.882	0.82	ξ_3 placed in [0.382, 1.0]
4	0.528	0.764	0.88	ξ_4 placed in [0.382, 0.764]
5	0.472	0.736	0.89	ξ_5 placed in [0.382, 0.618]
6	0.438	0.719	—	ξ_6 placed in [0.382, 0.528]

Thus, we first run the program providing the information: production level $x_1 = (0.691)10^6$. Then we obtain the prediction: profit, $p_1 \doteq \$(0.75)(10)^5$. This is recorded in the top row of the table. For $x_2 = (0.809)(10)^6$ we suppose a response of $p_2 = \$(0.87)(10)^5$.

Since the larger response is at $\xi = 0.618$, our next value is located in [0.382, 1.0]—in fact, at $\xi_3 = 0.764$ (as in the numerical example of Section 4.7). This yields $x_3 = (0.882)(10)^6$, and on feeding this into the black box, suppose that $p_3 = \$(0.82)(10)^5$ pops out. This implies our next ξ-point is to be located in [0.382, 0.764], namely, $\xi_4 = 0.528$.

Proceeding in this way, we suppose that the table above is compiled. We must stop at the ξ_6-value recorded there because this (together with ξ_5) violates the minimum spacing of points admitted, $\epsilon = 0.05$.

The model provided in the black box and the golden search technique therefore predict that the maximum profit will not be less than $(0.89)(10)^5 =$ $89,000$ and is to be attained at a production level determined by the ξ-interval $[0.382, 0.528]$. Since the accountants and management may not recognize what a "ξ-interval" is, we had better present the result in terms of an x-interval, namely, the maximum profit is to be attained at a production level between 691,000 and 764,000 Ping-Pong balls. The cost of our estimate is the cost of running the computer program six times.

Exercises for Section 4.7

1. Use trisection and then use a search by golden section to find an interval of uncertainty of width less than 0.1 for the maximum of

$$f(x) = -2x^2 + 3x + 1$$

   in $[0, 1]$. (Do not use calculus.)
2. Given that the function f, where $f(x) = 2(1 - x)\sqrt{x}$, is unimodal on $[0, 1]$, use four experiments to find an interval of uncertainty for the maximum of f in $[0, 1]$ by (a) dichotomous search, and (b) golden section. The minimum distance between experiments is to be 0.04. (Use calculations with about three significant figures. A table of square roots or, better still, a calculating machine will be helpful but not essential.)
3. Give formulas for the lengths, L_n, of the intervals of uncertainty after n function evaluations by equal spacing, trisection, and golden section, assuming the initial interval has length $L_0 = 2$. How large must n be before L_n (golden section) is less than one half of (a) L_n (equal spacing)? (b) L_n (trisection)?
4. The Fibonacci sequence is the set of numbers:

$$1, 1, 2, 3, 5, 8, 13, \ldots$$

   where each number is the sum of its two predecessors. Thus if F_n denotes the $(n + 1)$th member of the sequence,

$$F_n = F_{n-1} + F_{n-2} \qquad (n = 2, 3, \ldots)$$

   and $F_0 = F_1 = 1$.
   (a) Using Eq. (38) of Chapter 3 (justify its use), show that the Fibonacci numbers can be expressed in the form

$$F_n = \frac{1}{\sqrt{5}} (\tau_1^{n+1} - \tau_2^{n+1}) \qquad \text{for } n = 0, 1, 2, \ldots)$$

   where $\tau_1 = \frac{1}{2}(\sqrt{5} + 1)$, $\tau_2 = -\frac{1}{2}(\sqrt{5} - 1)$.
   (b) Prove that for any sequential search satisfying the relation (3), we have

$$L_{n-r} = F_{r-1}L_{n-1} + F_{r-2}L_n \qquad (\text{for } r = 2, 3, \ldots, n - 2).$$

Preliminary Exercises

Preliminary Exercises 1 and 2 of Chapter 2 (straight lines) are relevant for this chapter; Preliminary Exercises 5, 6 (geometric series), 8–13 (quadratic equations), and 14–16 (induction) of Chapter 3 are also relevant.

1. Let S be a set of real numbers. By $\max_{x \in S} (x)$ we mean a number a of S (if any) such that $a \geq b$ for all $b \in S$. If S is finite, say $S = \{a_1, a_2, \ldots, a_n\}$, then $\max (a_1, \ldots, a_n)$ is a number a_i in S such that $a_i \geq a_j$ for all a_j in S. A maximum of a *finite* set S always exists. For an infinite set S, the maximum may not exist, but it will always exist in our applications.

   Let

$$S_1 = \{5, -6, 3, -5, 2\}, \qquad S_2 = \{x \in R: 1 \leq x \leq 3\}.$$

$$S_3 = \{x \in R: 1 \leq x < 3\}, \qquad S_4 = \{x \in R: x = \frac{1}{1-y} \text{ and } 0 \leq y < 1\}$$

   Say whether $\max_{x \in S_i} (x)$ exists for $i = 1, 2, 3, 4$, and find this number when it does exist.

2. If f is a function defined on a set S of real numbers, then we define $\max f(x)$ to be $f(x_o)$, where x_o is a number of S (if any) for which $f(x_o) \geq f(x)$ for all $x \in S$.

   Find $\max_{x \in S} f(x)$, where it exists, if

   (a) $f(x) = x,$ $S = \{5, -6, 3, -5, 2\}.$
   (b) $f(x) = |x|,$ $S = \{5, -6, 3, -5, 2\}.$
   (c) $f(x) = 1/x,$ $S = \{x \in R: 0 < x \leq 1\}.$
   (d) $f(x) = 1,$ $S = R.$
   (e) $f(x) = \begin{cases} \frac{1}{2} & \text{(if } 0 \leq x < \frac{1}{2}), \\ 2(1 - x) & \text{(if } \frac{1}{2} \leq x \leq 1), \end{cases}$ $S = \{x \in R: 0 \leq x \leq 1\}.$
   (f) $f(x) = x(1 - x),$ $S = R.$

3. (a) Show that

$$4x^2 - 9y^2 = (2x + 3y)(2x - 3y)$$

$$2ax - 4bx + ay - 2by = (a - 2b)(2x + y).$$

   (b) Find the real factors of lowest degree of:

$$2z^3 + 10z^2 - 28z, \qquad x^2 - y^2$$

$$x^3 - y^3, \qquad x^3 - xy^2 - xy^2 + y^3.$$

4. If $a \leq m$ and $b \geq M$, show that $a/b \leq m/M.$

5. (Check Preliminary Exercise 1 of Chapter 3 before beginning this exercise.)
   (a) Let b be a fixed positive real number. Sketch graphs of b^x for all real x when (i) $b = \frac{1}{4}$, (ii) $b = 1$, and (iii) $b = 4$.

(b) If N is a positive real number and $b \neq 1$, there is one and only one number x such that $b^x = N$. This number x is the logarithm of N to the base b, and we write $x = \log_b N$. Use part (iii) of 5 (a) as a guide to sketch a graph of the function f if $f(y) = \log_4 y$ for all $y > 0$. Evaluate $\log_3 9$, $\log_2 8$.

(c) A law of exponents says that, for any $x \geq 0$, $y \geq 0$,

$$b^x b^x = b^{x+y}.$$

By writing $x = \log_b M$ and $y = \log_b N$, use this to show that $\log_b (MN) = \log_b M + \log_b N$.

(d) A second law of exponents implies that $1/b^x = b^{-x}$. By writing $x = \log_b M$ and observing $1/M = 1/b^x = b^{-x}$, show that

$$\log_b \left(\frac{1}{M}\right) = -\log_b M$$

and hence that, if $M > 0$, $N > 0$, then

$$\log_b \left(\frac{M}{N}\right) = \log_b M - \log_b N.$$

6. Let f be a real-valued function defined on a set S of real numbers. We say that f is *increasing on* S if $x_1 < x_2$ and $x_1, x_2 \in S$ imply $f(x_1) < f(x_2)$. Let f be defined by $f(x) = x^2$. Is f an increasing function on S if (a) $S = \{x \in R : x \geq 0\}$, and (b) $S = R$? If f is defined by $f(x) = \log_b x$ (see Preliminary Exercise 5) and $S = \{x \in R : x > 0\}$, show that f is an increasing function on S.

CHAPTER FIVE

POPULATION GENETICS

5.1 Introduction

One of the more fascinating and, at one time, controversial developments of modern science concerns studies of the evolution of plant and animal species. We are now going to develop some simple models which relate to the propagation of physical characteristics through the successive generations of a species. These models will, however, have a feature that is not shared by any of the models used in the other chapters of this book. The distinction to be made is between "deterministic" models, as used elsewhere, and "probabilistic" models as used in this chapter.

In any of the models introduced so far, we have been able to predict some conclusion with mathematical certainty, given all the axioms and hypotheses which went into the model. We shall now investigate models which make predictions only in the form of probabilities. For example, we might be able to make an assertion such as: "If, in a particular human population, both parents of a child have red hair, then there is an even chance that the child will have red hair." The notion of probability is contained in the phrase "even chance." Before we can begin with our model building, it is necessary for us to clarify the idea of probability. Our discussion will be confined to those concepts needed for the models of this chapter, and no attempt is made to go beyond these immediate needs. The more thorough and complete accounts given by Goldberg (1960) or Derman, Glaser, and Olkin (1973) can be recommended.

5.2 Probability

We often use the ideas of chance, likelihood, or probability in everyday language. For example, "It is unlikely to rain today," "The black horse will probably win the next race," or "A playing card selected at random from a pack is unlikely to be the ace of spades." Each of these remarks, if accepted at face value, is likely to reflect the speaker's expectation based on experience gained in the same position, or similar positions, on many previous occasions. In order to be quantitative about probability, we focus on this aspect of *repeatable* situations.

Since we are working in the mathematical realm of ideas, we are at liberty to consider the ideal situation of a procedure that is to be followed under conditions which do not vary from one performance to the next. In these circumstances, we say that we are performing an *experiment*. For example, the act of drawing a playing card at random from a pack of 52 such cards is readily accepted as fitting our description of an experiment. It is harder to fit a weather prediction into this framework, but even in this case, we can expect the first statement quoted above to be based on the weather observed to have developed from the same weather indicators on previous occasions.

A further demand we must make of an experiment is that all possible conclusions, or outcomes, can be described. The set of all possible outcomes is called the *sample space* for the experiment. Thus in drawing the playing card, the sample space is a set with 52 members. The experiment of spinning a coin is generally assumed to have two members in the sample space labeled as "heads" and "tails". The procedure, or experiment, of tossing the coin twice gives rise to a sample space represented by $\{HH, HT, TH, TT\}$. If we consider an urn containing a mixture of ten red, five black, and twenty green balls and perform the experiment of selecting a ball at random, the sample space consists of three members corresponding to the selection of a ball of each of the three colors. It could be represented by $\{R, B, G\}$. We shall always confine our attention to experiments in which the sample space is a *finite* set.

In each of the situations described, we are tempted to think in quantitative terms about the possible outcomes. Thus, there is 1 chance in 52 of drawing the ace of spades, or 5 chances in 35 of picking a black ball, and so on. We formalize these ideas in the following way. To each outcome represented in the sample space, we attach a number in the interval [0, 1] called the *probability* of that outcome. However, the numbers must also satisfy the condition that the sum of all the probabilities of all outcomes in the sample space is 1. Thus, if $\{A_1, \ldots, A_n\}$ represents the sample space, we write $P(A_i)$ for the probability of outcome A_i, and then

$$0 \leq P(A_i) \leq 1 \qquad (\text{for } i = 1, 2, \ldots, n)$$

and

$$P(A_1) + \cdots + P(A_n) = 1.$$

These probabilities can, in theory, be assigned quite arbitrarily within the above rules. However, in any given experiment, the assignment is guided by intuition or measurement. For example, since we consider it equally likely that any playing card will be drawn from a pack, the probability attached to each outcome had better be the same. We are led to a probability of $\frac{1}{52}$ for drawing the ace of spades as, indeed, for any other specified card. In the urn problem, we naturally attach probabilities of $\frac{10}{35} = \frac{2}{7}, \frac{5}{35} = \frac{1}{7}$, and $\frac{20}{35} = \frac{4}{7}$ to R, B, and G, respectively.

Suppose now that we have an experiment with sample space $\{A_1, A_2, \ldots, A_n\}$. We frequently wish to attach a probability to the outcome being one of several members of the sample space. For example, we may wish to say something about the probability that a playing card drawn at random is an ace. In mathematical terms, we are to find a mechanism for attaching a probability to *subsets* of the sample space. For convenience, such a subset is described as an *event*. The very natural rule, which we accept as an axiom, is this: The probability of any event is the sum of the probabilities of the outcomes which make up the event. An event is said to *occur* if the outcome of the experiment is in the event.

To return to the playing cards again, if AH, AC, AD, AS denote the outcomes ace of hearts, ace of clubs, etc., then the event E is given by

$$E = \{AH, AC, AD, AS\}$$

and

$$P(E) = P(AH) + P(AC) + P(AD) + P(AS)$$
$$= \tfrac{1}{52} + \tfrac{1}{52} + \tfrac{1}{52} + \tfrac{1}{52} = \tfrac{1}{13}.$$

In more colloquial language, if the event consists of outcomes $A_1, A_2, \ldots, A_k$ (i.e., $E = \{A_1, A_2, \ldots, A_k\}$), then $P(E)$ is the probability that the outcome of an experiment is *either* A_1, or $A_2, \ldots$, or A_k. The axiom then yields

$$P(E) = \sum_{i=1}^{k} P(A_i).$$

For the purposes of the models to be built up in this chapter, we need just two rules for handling the probabilities of compound events. The first can readily be verified by our definition for the probability of an event and the definitions of the union and intersection of pairs of sets (see Exercise 3.).

Rule 1. If E_1 and E_2 are events (they may or may not be in the same sample space), then

$$P(E_1 \cup E_2) = P(E_1) + P(E_2) - P(E_1 \cap E_2).$$

Observe that the event $E_1 \cup E_2$ occurs if the outcome of an experiment is in either of E_1 or E_2; i.e. if and only if either E_1 or E_2 occurs. For informal discussion, we keep in mind the intuitive statement: $E_1 \cup E_2$ is the event consistent with *either E_1 or E_2*. If E_1, E_2 are events from different sample spaces of separate experiments, then, obviously, $E_1 \cap E_2$ is empty and $P(E_1 \cup E_2) = P(E_1) + P(E_2)$.

The second rule rests on a more subtle idea of probability theory; namely, that of *statistical independence*. Intuitively, experiments are thought to be independent if the outcome of one experiment can have no influence on the outcome of the other. Thus, the probabilities associated with the sample space of one experiment are not influenced in any way by the conclusion of the other experiment. It is sometimes the case that the experiments in question are repetitions (or *trials*) of the same experiment. For example, a die may be rolled twice. Or, what is equivalent for our purposes, two dice may be rolled simultaneously. These are obviously independent experiments. If we repeat the experiment of selecting a card from a well-shuffled pack of playing cards a similar situation occurs. But note that the card selected in the first trial must be replaced for the second trial if the same pack is to be used. Otherwise the second trial will constitute a different experiment from the first and the trials will not be independent in the sense described; without replacement the members of the sample space in the second experiment depend in an obvious way on the outcome of the first experiment. It is often convenient to think of experiments as being performed in sequence even though, in real life, they may be performed simultaneously.

For a deeper discussion of independence the reader is referred to Goldberg (1960) and Derman et al. (1973) of the bibliography or to any other modern text on probability theory.

Rule 2. If E, F are events in the sample spaces of two independent experiments then the probability that E occurs in the first *and* F occurs in the second is $P(E) \cdot P(F)$.

For informal discussion, we abbreviate and say $P(E \text{ and } F) = P(E)P(F)$ but, in use, we must always keep in mind the proviso that E, F are from independent experiments. To illustrate, suppose that a die is rolled and E is the event that an even number occurs. Then the sample space is $\{1, 2, 3, 4, 5, 6\}$ each event having probability $\frac{1}{6}$ and $E = \{2, 4, 6\}$ with $P(E) = \frac{1}{2}$. The second experiment is to select a card at random from a pack of 52 playing cards and F is the event that a diamond is selected. Then the sample space has 52 members each with probability $\frac{1}{52}$ and F is a subset of 13 members from the sample space so that $P(F) = \frac{1}{4}$. These are obviously independent experiments. Rule 2 says that the probability that one gets an even number on the die and also gets a diamond is $P(E) \cdot P(F) = (\frac{1}{2}) \cdot (\frac{1}{4}) = \frac{1}{8}$.

The next example should be studied in detail as it illustrates some important ideas which will recur in our analysis of problems in genetics.

EXAMPLE. Suppose that there are two urns, each containing twenty balls. In urn 1 there are three black, seven green, and ten red balls, while urn 2 contains six black, ten green, and four red balls. One ball is selected from each of urns 1 and 2 in turn. What are the probabilities of obtaining (a) two black balls, (b) one black and one green ball, and (c) *either* a green ball from urn 1 or a black ball and a green ball?

For part (a) we view the two selections as independent experiments. The sample spaces are denoted by $\{b_1, g_1, r_1\}$ and $\{b_2, g_2, r_2\}$ for the selection from urns 1 and 2, respectively. If $E_1 = \{b_1\}$ and $E_2 = \{b_2\}$, we are asked for $P(E_1$ and $E_2)$. It is clear that E_1 and E_2 are independent and, since $P(E_1) = \frac{3}{20}$ and $P(E_2) = \frac{6}{20}$, rule 2 yields

$$P(E_1 \text{ and } E_2) = \tfrac{3}{20} \cdot \tfrac{6}{20} = \tfrac{9}{200}.$$

In part (b) we consider the two selections to constitute one trial of the experiment. The possible outcomes are then pairs which may be listed as follows:

$$\{(b_1, b_2), (b_1, g_2), (b_1, r_2), (g_1, b_2), (g_1, g_2), (g_1, r_2), (r_1, b_2), (r_1, g_2), (r_1, r_2)\}.$$

This is now the sample space, and using the argument of part (a), the associated probabilities are found to be

$$\{\tfrac{9}{200}, \tfrac{3}{40}, \tfrac{3}{100}, \tfrac{21}{200}, \tfrac{7}{40}, \tfrac{7}{100}, \tfrac{3}{20}, \tfrac{1}{4}, \tfrac{1}{10}\}$$

The outcome of one black and one green ball can be achieved in two ways:

$$(b_1, g_2) \quad \text{or} \quad (g_1, b_2).$$

Thus, the event in question is

$$E = \{(b_1, g_2), (g_1, b_2)\},$$

and from the definition of the probability of an event,

$$P(E) = P(b_1, g_2) + P(g_1, b_2) = \tfrac{3}{40} + \tfrac{21}{200} = \tfrac{9}{50}.$$

For part (c) let F be the event that the first ball drawn is green. Then, in terms of the above sample space, we have

$$F = \{(g_1, b_2), (g_1, g_2), (g_1, r_2)\} \quad \text{and} \quad P(F) = \tfrac{7}{20}.$$

With E as defined in part (b), we are now asked to find the probability of either E or F. In other words, $P(E \cup F)$. By rule 1 we have

$$P(E \cup F) = P(E) + P(F) - P(E \cap F).$$

Since $E \cap F = \{(g_1, b_2)\}$, $P(E \cap F) = \frac{21}{200}$. Thus,

$$P(E \cup F) = \frac{9}{30} + \frac{7}{20} - \frac{21}{200} = \frac{85}{200} = \frac{17}{40}.$$

Exercises for Section 5.2

1. For each of the following experiments, specify a sample space:
   (a) A card is selected at random from a pack of cards.
   (b) Three coins are tossed (taking account of order).
   (c) A telephone number is chosen at random from the directory, and the person who answers is asked whether he or she is listening to a particular radio station.
   (d) The coats of mice may be described as agouti or nonagouti. Litters of exactly three mice are observed and the occurrence of agouti coats among the three are noted.
   (e) In case (d), the sex of each mouse is to be recorded as well as the nature of its coat.
   (f) A survey of families having exactly two children (not twins) is made, and their sexes are to be recorded with the older child first.

2. (a) For case 1(f) above, describe (as a set) the event in which the first-born child is a girl.
   (b) For case 1(d) above, describe (as a set) the event in which exactly two mice of the litter have agouti coats.
   (c) Repeat step 2(b), but for the experiment (e) of Exercise 1.

3. Establish rule 1 in Section 5.2. (*Hint:* Let E_1 and E_2 have outcomes $A_1, \ldots, A_p$ in common, and only these. Then write

$$E_1 = \{A_1, \ldots, A_p, B_1, \ldots, B_q\}, \qquad E_2 = \{A_1, \ldots, A_p, C_1, \ldots, C_r\}$$

and we have

$$E_1 \cup E_2 = \{A_1, \ldots, A_p, B_1, \ldots, B_q, C_1, \ldots, C_r\},$$
$$E_1 \cap E_2 = \{A_1, \ldots, A_p\}.)$$

4. Let A be the sample space of an experiment and let E be an event (a subset of A). If E^c (the *complement* of E) denotes the event consisting of all outcomes in A excluding those in E, prove that

$$P(E) = 1 - P(E^c)$$

5. If a single die is thrown, what is the probability that
   (a) An odd number turns up?
   (b) A number greater than two turns up?

6. If two dice are thrown, what is the probability that
   (a) Two sixes turn up?
   (b) The sum of the two numbers turned up is three?
   (c) The sum of the two numbers turned up exceeds seven?
7. Four married couples are seated at random at a round table. What is the probability that a particular husband and wife will sit together?
8. An experiment consists of finding a randomly chosen person's sex and whether he/she has blue eyes. (Assume one can always decide!). Let F, B denote the events that a person is female, has blue eyes, respectively. What is the sample space? Express in terms of $P(F), P(B)$:
   (a) The probability of selecting a male with blue eyes.
   (b) The probability of selecting a male who does not have blue eyes.
9. (a) In a cross of certain black (hybrid) with white (recessive) guinea pigs, it is known that each offspring is black (event B) or white (event W) with equal probability and that $P(B) = P(W) = \frac{1}{2}$. In a litter of two offspring, what is the probability that one pig is white and the other black? In a litter of three, what is the probability that two are black and one is white?
   (b) If both parents are hybrid, then it is known that $P(B) = \frac{3}{4}$ and $P(W) = \frac{1}{4}$. Answer the questions of part (a) in this case.
10. (a) For a certain type of aircraft engine and a certain journey, the probability that an engine will *not* fail is p. What is the probability that an aircraft flying on two such engines will have them both fail?
   (b) If an aircraft with four such engines is able to fly on three engines but not on two, show that it is always safer to fly on the two-engine aircraft.
   (c) Investigate the case in which the four-engine aircraft can fly on 4, 3, or 2 engines.

5.3   Genetics

For our purposes, a *gene* is a basic biological unit of molecular proportions which is found in each individual of a living species. The genes contain coded information which determines

1. the living cells which must be continually produced as long as the individual is alive, and
2. the cells, and hence the physical characteristics, of the individual's descendants.

Throughout this chapter, our model of a gene will be primitive from the biochemical point of view. Thus, we shall first assume that a gene is a unit whose composition we ignore, which is not affected by changes in its environment, and which is transmitted without change of form from one generation to the next. These hypotheses are an integral part of the process of making our first mathematical model to investigate the phenomenon of inherited characteristics.

In many cases, experiment and analysis indicate that the form of certain physical characteristics of living creatures is dictated, not by single genes, but by groups of genes. In particular, we are going to investigate physical characteristics which are determined by genes which (a) themselves belong to one of two types and (b) always occur in pairs.

The two gene types are called *dominant* (D) and *recessive* (R); therefore the possible pairs of genes (ordering is not relevant) can be labeled DD, RD, and RR. Now all the gene pairs carried by one individual, and associated with a particular physical feature, are all of the same type. For example, in humans the characteristic of eye color, brown (D) or blue (R), can be associated with a gene pair. Individuals with brown eyes have the governing gene pairs which are DD or RD, and blue eyes indicate gene pairs of type RR.

Thus, we confine our modeling to those inherited characteristics appearing in a population which may be identified with the three kinds of gene pairs. The population may then be decomposed into three mutually exclusive subsets, or *genotypes*:

1. DD, which is described as *dominant* or *homozygous dominant*.
2. RD, which is described as *hybrid* or *heterozygous*.
3. RR, which is described as *recessive* or *homozygous recessive*.

Now we focus attention on a particular generation and suppose that, of the total population in that generation, the *proportions* of the three genotypes present are d, $2h$, and r, respectively.* Then d, $2h$, and r are nonnegative numbers with

(1) $$d + 2h + r = 1.$$

Furthermore, if there are n members of the population, then the numbers of dominant, hybrid, and recessive genotypes are dn, $2hn$, and rn.

We can phrase this last idea in a different way using the notion of probability introduced in the preceding section. It would be equivalent to say that, in a random selection of one individual from the population, the probability of selecting a dominant, hybrid, or recessive is d, $2h$, or r, respectively. Since every individual is of one and only one genotype, we again arrive at Eq. (1).

If the reader finds this model building rather complicated, comfort can be found in the fact that the idea of classifying individuals into genotypes took many years to evolve and was created as a device to explain a lot of accumulated experimental evidence. As in other parts of this book, we have to omit any detailed account of the critical experiments and historical evolution of the concepts.

*We use $2h$, rather than h, only for our subsequent convenience.

The next phase of our modeling is to consider in more detail the transmission of genotypes from one generation to another, and in doing so, we must formulate axioms which will describe mathematically a mechanism of inheritance. We first formalize the idea of a generation and assume that the population of interest can be described by a sequence of *discrete* generations. This means that all individuals survive for just one generation and that there is no overlap between successive generations. This is by no means obvious for the human species, for example, but it is a sound hypothesis in describing annual crops of wheat.

Suppose that the generations are labeled 0, 1, 2, ... and that, in the nth generation, the proportions of the three genotypes present are d_n, $2h_n$, and r_n, in an obvious notation. Then for $n = 0, 1, 2, \ldots$ we have

$$(2) \qquad\qquad d_n + 2h_n + r_n = 1.$$

We assume next that the population is *bisexual* and that the proportions of the genotypes are identical in the population of males and in the population of females. The genotype of an individual in generation $n + 1$ is then supposed to be determined by the genotypes of its two parents in generation n. In a typical gene pair of the individual, one gene is inherited from the male parent and is *selected at random* from the two genes of the male parent. Similarly, the second gene of the pair is inherited from the female parent and is selected at random from the two genes carried by this parent.

If both parents are homozygous (either dominant or recessive), it is clear that the selections "at random" are meaningless, and the genotype of the offspring is determined with certainty (probability 1) as follows:

Genotypes of Parents	Genotype of Offspring
DD and *DD*	*DD*
DD and *RR*	*RD*
RR and *DD*	*RD*
RR and *RR*	*RR*

If at least one of the parents is a hybrid, then the genotype of the offspring cannot be predicted with certainty. For example, if one parent is *DD* and the other *RD*, then one gene is certainly *D* and the other is either *R* or *D*, each outcome being associated with a probability of $\frac{1}{2}$. This is because the genes of the hybrid are assumed to be divided evenly between dominants and recessives, and we select at random from this gene population. Thus, in this case, the offspring is of genotype *DD* or *RD*, each outcome having a probability $\frac{1}{2}$.

Where both parents are hybrid, the offspring may be dominant, hybrid, or recessive. To find the probabilities of each of these outcomes, we must use the basic rules of probability discussed in the preceding section. For the moment, describe the possible outcomes of selecting a gene from the male or female parent as $\{D_m, R_m\}$ and $\{D_f, R_f\}$, respectively and for both experiments, each outcome has probability a half. We now consider the compound experiment consisting of:

1. Selecting a gene at random from the two carried by the male parent.
2. Selecting a gene at random from the two carried by the female parent.

The sample space is

$$\{D_m \text{ and } D_f, D_m \text{ and } R_f, R_m \text{ and } D_f, R_m \text{ and } R_f\}.$$

Using rule 2 we associate with each outcome a probability of $\frac{1}{4}$. For example $P(D_m \text{ and } D_f) = P(D_m)P(D_f) = \frac{1}{4}$. Now it is clear that the occurrence of a dominant, hybrid, or recessive offspring corresponds, respectively, to the events $\{D_m \text{ and } D_f\}, \{D_m \text{ and } R_f, R_m \text{ and } D_f\}, \{R_m \text{ and } R_f\}$. The associated probabilities are therefore:

$$P(\text{dominant}) = \frac{1}{4}, \quad P(\text{hybrid}) = \frac{1}{2}, \quad P(\text{recessive}) = \frac{1}{4}.$$

In the following table, we indicate the probabilities of each of the three genotypes, in the order DD, RD, RR, arising from each possible pairing of parent genotypes.

TABLE 5.1

Probabilities of Offspring Genotypes.

Parent 1 \ Parent 2	DD	RD	RR
DD	$(1, 0, 0)$	$(\frac{1}{2}, \frac{1}{2}, 0)$	$(0, 1, 0)$
RD	$(\frac{1}{2}, \frac{1}{2}, 0)$	$(\frac{1}{4}, \frac{1}{2}, \frac{1}{4})$	$(0, \frac{1}{2}, \frac{1}{2})$
RR	$(0, 1, 0)$	$(0, \frac{1}{2}, \frac{1}{2})$	$(0, 0, 1)$

Exercises for Section 5.3

1. Consider a physical trait in which the recessive form is lethal, and let d, h, r have the meanings used in formulating Eq. (1). What are the values of d, h, r when the highest possible number of recessive genes is present in the population? What is the mortality rate (due to the birth of recessives) in this case?
2. Suppose that in a bisexual population as described in this section, there is a probability α that, after a dominant gene is selected from a male parent (but not

a female parent), the gene will mutate to a recessive gene on transmission to the offspring. For example, if both parents are of dominant genotype, then the offspring is dominant with probability $1 - \alpha$, hybrid with probability α, and recessive with probability 0. Thus, the entry in the top left corner of Table 5.1 is replaced by $(1 - \alpha, \alpha, 0)$. Complete the corresponding table in this case. (Note that, when $\alpha = 0$, Table 5.1 should be recovered.)

5.4 Population Genetics

From the point of view of evolution, our interest focuses on the transmission of physical characteristics through large populations. For example, if the size of a head of corn is governed by a gene-pair mechanism as we have described it, and if one crop is always used to seed the next, then a farmer would be interested in knowing whether this characteristic relating to yield is going to become more or less predominant with each generation. For the human race, we may ask whether the phenomenon of blue eyes is becoming more or less frequent as the race evolves. In the discussion of Section 5.3, we described a model for the transmission of genotypes from a given pair of parents. If we are examining a large population, we cannot examine the genotypes of each such pair in order to make predictions about the whole of the next generation. Consequently, we introduce more hypotheses, of a probabilistic type, to treat these cases.

We assume that (typically) an individual in generation $n + 1$ is formed in the following way:

1. Male and female parents are each selected at random from the populations of males and females in generation n.
2. The genotype of the offspring is determined by selecting a gene at random from (a) those carried by the male parent and (b) those carried by the female parent.

These rules describe the process known as *panmixia*, or *random Mendelian mating*. Note that, given the parent genotypes, the second assumption is just the model of inherited genotypes discussed in Section 5.3.

The problem is now the following: Given the frequencies d_n, $2h_n$, and r_n of dominants, hybrids, and recessives in the nth generation, and assuming random Mendelian mating, what are the corresponding frequencies, d_{n+1}, $2h_{n+1}$, and r_{n+1} in the $(n + 1)$th generation?

A randomly selected individual in generation $n + 1$ has two genes; one from each parent. The experiment of selecting such an individual at random and identifying the gene inherited from each parent (in generation n) has the sample space:

$$\{D_m \text{ and } D_f, \; D_m \text{ and } R_f, \; R_m \text{ and } D_f, \; R_m \text{ and } R_f\}$$

where D, R denote the type of gene and the subscript says from which parent the gene is inherited. Then we see that

(3)
$$d_{n+1} = P(D_m \text{ and } D_f), \quad 2h_{n+1} = P(D_m \text{ and } R_f) + P(R_m \text{ and } D_f),$$
$$r_{n+1} = P(R_m \text{ and } R_f)$$

so that our problem is solved if we can express these four probabilities in terms of d_n, h_n and r_n.

First consider D_m and R_m. These are events in the sample space for the compound experiment: Select a male individual in generation n and then select a gene from the pair carried by this individual. The sample space can be written

(4)
$$\{\delta_m D, \gamma_m D, \gamma_m R, \rho_m R\}$$

where the Greek letters δ, γ, ρ, denote the genotype of the individual (dominent, hybrid, or recessive respectively) and the D, R denote the subsequent selection of a dominant or recessive gene. The probabilities associated with the first and last outcomes are obviously d_n, r_n respectively. Since $d_n + 2h_n + r_n = 1$, and the sum of the probabilities of the four outcomes must be one, this leaves a probability of $2h_n$ for the event $\{\gamma_m D, \gamma_m R\}$ and we divide it evenly between these two outcomes. Thus the four outcomes in this sample space have associated probabilities: d_n, h_n, h_n, r_n.

Since $D_m = \{\delta_m D, \gamma_m D\}$ and $R_m = \{\gamma_m R, \rho_m R\}$ we now see that $P(D_m) = d_n + h_n$ and $P(R_m) = h_n + r_n$.

Turning to the events D_f and R_f we can go through exactly the same analysis as for D_m and R_m simply replacing "male" by "female" everywhere. Since the genotypes are distributed in the male and female populations in just the same way as in the total population it is found that $P(D_f) = d_n + h_n$ and $P(R_f) = h_n + r_n$.

Finally, since the experiments of selection from male and female populations are obviously independent, we may apply rule 2 of section 5.2 in Eqs. (3) to find

$$d_{n+1} = P(D_m \text{ and } D_f) = P(D_m)P(D_f) = (d_n + h_n)^2$$
$$2h_{n+1} = P(D_m)P(R_f) + P(R_m)P(D_f) = 2(d_n + h_n)(h_n + r_n)$$
$$r_{n+1} = P(R_m)P(R_f) = (h_n + r_n)^2.$$

We have now completed the major part of the analysis. It remains to examine the significance of the nonlinear recurrence relations:

(5)
$$d_{n+1} = (d_n + h_n)^2$$
$$2h_{n+1} = 2(d_n + h_n)(h_n + r_n)$$
$$r_{n+1} = (h_n + r_n)^2.$$

But first let us remark that we have met nonlinear recurrence relations before; Eqs. (22) and (23) of Chapter 3, for example. In both cases we will use the relation to study the evolution of certain phenomena with time; in this case, the object of interest is the variation in the frequencies (proportions) of the three genotypes with different generations.

The implications of the relations (5) are striking and remarkably simple. To see something of the nature of sequences $\{d_0, d_1, \ldots\}$, $\{h_0, h_1, \ldots\}$, and $\{r_0, r_1, \ldots\}$ which satisfy these relations, we shall proceed one step further and examine the genotype frequencies predicted for generation $n + 2$. We first introduce the abbreviations

(6)
$$p_n = d_n + h_n, \qquad q_n = h_n + r_n.$$

The Eq. (2) implies $p_n + q_n = d_n + 2h_n + r_n = 1$, and the recurrence relations become

(7)
$$d_{n+1} = p_n^2, \qquad 2h_{n+1} = 2p_n q_n, \qquad r_{n+1} = q_n^2.$$

These relations are, of course, valid for any positive integer n. In particular, we have from the first of Eqs. (5):

$$d_{n+2} = (d_{n+1} + h_{n+1})^2$$
$$= (p_n^2 + p_n q_n)^2 = p_n^2 (p_n + q_n)^2 = p_n^2,$$

since $p_n + q_n = 1$. Thus, we find that $d_{n+2} = d_{n+1}$ and the proportion of the dominant genotype is *unchanged* from generation $n + 1$ to generation $n + 2$.

It can easily be confirmed that we also have

$$h_{n+2} = h_{n+1} \quad \text{and} \quad r_{n+2} = r_{n+1}.$$

Now it is clear that we can repeat the argument indefinitely and conclude that, for every $k \geq n + 1$,

$$d_k = d_{n+1}, \qquad 2h_k = 2h_{n+1}, \qquad r_k = r_{n+1}.$$

Thus our model predicts that, whatever the proportions of genotypes in generation n, they will in all subsequent generations maintain those proportions attained in generation $n + 1$. This is the celebrated *Hardy-Weinberg* law and, in the terminology of Chapter 3, it indicates the very rapid development of an *equilibrium* genotypic distribution. This prediction is, of course, subject to all the hypotheses made in the construction of our model and generally implied by the term panmixia.

The phenomenon predicted is, however, of the greatest physical significance. It is a matter of observation, and taken very much for granted, that the presence of physical characteristics in a population is remarkably stable. If there were to be complete stability and *no* changes in physical characteristics, then there would be no evolution. Indeed, evolution may be described as a trend in physical characteristics away from some apparently stable equilibrium type. Thus, the Hardy-Weinberg situation is important because it describes situations in which a high degree of stability is associated with the equilibrium configuration. We may then hope to obtain viable models of evolving populations by postulating some changes in the model which will retain the major features and yet predict some slow variation in genotype frequencies as the generations progress.

We conclude this section with a summary of the hypotheses which make up our model of inheritance of a particular physical characteristic:

1. The controlling genes exist in two types, D and R.
2. Individuals carry genes in pairs giving rise to individuals of three genotypes.
3. The population is bisexual and has the same distribution of genotypes in both sexes.
4. Generations are discrete.
5. Items 1 and 2 of page 114.

Exercises for Section 5.4

1. Verify from the recurrence relations (5) that

$$d_{n+1} + 2h_{n+1} + r_{n+1} = 1.$$

2. From Eqs. (6) and (7) we see that, for a population in equilibrium under panmixia, we have

$$d = p^2, \qquad 2h = 2pq, \qquad r = q^2$$

where $0 \leq p, q \leq 1$ and $p + q = 1$. Sketch a graph of the equation $p + q = 1$ and (on the same sketch) $pq = h$ for three or four values of h chosen from the interval $[0, 1]$. Use this to convince yourself that the maximum possible proportion of hybrids in such an equilibrium population is $\frac{1}{2}$ and that, in this case, $d = \frac{1}{4}$, $2h = \frac{1}{2}$, $q = \frac{1}{4}$.

3. Show that, in a population in equilibrium under panmixia,

$$h = \sqrt{dr}.$$

4. The proportions of the three genotypes in a population with panmixia are $(0.16, 0.48, 0.36)$. By calculating the proportions in the next generation, show that this is an equilibrium distribution. *Note:* In this and subsequent exercises, a triple of real numbers written as above denotes proportions of dominants, hybrids, and recessives, respectively.

5. Find the proportions of genotypes in the next generation for each of the following distributions:

$$(0.2, 0.2, 0.6), \qquad (0.3, 0, 0.7), \qquad (0, 0.6, 0.4).$$

6. Decide which of the following genotype frequencies correspond to equilibrium distributions under panmixia. Where they do not, find the equilibrium frequencies.

$$(0.1, 0.2, 0.7), \qquad (0.5, 0, 0.5), \qquad (0.4, 0.4, 0.2), \qquad (\tfrac{25}{36}, \tfrac{10}{36}, \tfrac{1}{36}).$$

7. In a population in equilibrium under panmixia, five percent are observed to be recessives. What proportion are hybrids?

8. A survey of 14,345 foxes yielded 13,655 with red coats, 12 with black coats, and 678 with coats of intermediate coloration. Are these figures consistent with an equilibrium distribution of genotypes under panmixia?

9. (Cf. Exercise 1 of Section 5.3, and Exercise 4 of Section 5.6.) Consider a physical trait in which the recessive form is lethal. Show that, with d_n, h_n, and r_n as defined above, under the conditions of panmixia, the mortality rate due to the birth of recessives is h_n^2. Prove also that the genotype frequencies satisfy the recurrence relations

$$d_{n+1} = \frac{d_n + h_n}{1 + h_n}, \qquad 2h_{n+1} = \frac{2h_n}{1 + h_n} \qquad (n = 0, 1, 2, \ldots).$$

(Note that the frequencies in generation $n + 1$ are related only to those which *survive* in that generation.)

5.5 Breeding by Selection

In order to show that, in extreme situations at least, certain physical characteristics *can* become more and more pervasive with succeeding generations, we shall radically change the panmixia model. In Section 5.3, we introduced the idea of a bisexual population and assumed there (and subsequently to this point) that "the proportions of genotypes are identical in the population of males and the population of females". We now imagine some agency which admits only males of the dominant genotype to breed. Intuitively, we would then expect the subsequent generations to have more and more dominant genotypes.

In real life, this kind of model may be appropriate in the study of crops of corn, of petunia flowers, or sable farms, for example. In such cases, the dominant gene is likely to characterize a favorable physical characteristic—a larger corn cob, a prettier flower, or a thicker pelt. For both males and females, we continue to assume that the populations are large; indeed, we shall retain all the features of the panmixia model with the one exception noted above. It should be remarked that, if only a small number of males are selected for breeding, there is some danger of weakness in the subsequent populations due to inbreeding—a phenomenon which we do not attempt to represent in our model.

We now proceed to formulate the mathematical conditions implied by our new model. Suppose that, in the first generation in which only dominant genotypes are selected from the males for breeding, the proportions of genotypes in the female population are $d_0, 2h_0, r_0$, in a notation consistent with the earlier analysis.

The sample space associated with the genotype of an offspring now contains only two outcomes:

$$\{D_m \text{ and } D_f, \; D_m \text{ and } R_f\}.$$

Since we are associating probability one with D_m and, as in the preceding section, $P(D_f) = d_0 + h_0$, $P(R_f) = h_0 + r_0$, we obtain

$$P(D_m \text{ and } D_f) = P(D_f) = d_0 + h_0, \quad P(D_m \text{ and } R_f) = P(R_f) = h_0 + r_0.$$

Thus

(8)
$$\begin{cases} d_1 = d_0 + h_0 \\ 2h_1 = h_0 + r_0 \\ r_1 = 0. \end{cases}$$

It is clear that no recessives will appear in any subsequent generation. Thus, in generation n, with $n \geq 1$, we have $r_n = 0$. The proportions of genotypes in generation $n + 1$ are then obtained by comparison with (8) in the form

(9)
$$\begin{aligned} d_{n+1} &= d_n + h_n \\ 2h_{n+1} &= h_n \end{aligned} \qquad (n = 1, 2, 3, \ldots).$$

Supposing that $d_0, 2h_0, r_0$ are known, the relations (9) together with the initial conditions (8) determine a pair of simultaneous recurrence relations. Unlike the situation arrived at in the preceding section (Eq. (5)), these relations are *linear*. Using the methods developed in Chapter 3, this problem can be solved very easily to give $d_{n+1}, 2h_{n+1}$ explicitly in terms of $d_0, 2h_0$, and r_0. We leave it as an exercise for the reader to complete stage 3 of the modeling process (see Chapter 1) and prove that, for $n = 0, 1, 2, \ldots$,

(10)
$$d_{n+1} = d_0 + \left(1 - \frac{1}{2^{n+1}}\right)(2h_0) + \left(1 - \frac{1}{2^n}\right)r_0$$

$$2h_{n+1} = \frac{1}{2^{n+1}}(2h_0) + \frac{1}{2^n}r_0 = \frac{1}{2^n}(h_0 + r_0).$$

It is clear from this solution that, as n increases, d_{n+1} approaches the value $d_0 + 2h_0 + r_0 = 1$ (by Eq. (1)), and h_{n+1} approaches the value 0. Thus, as anticipated, recessives disappear in the first generation, the proportion of hybrids gets less and less (being halved in successive generations after the first), and the proportion of dominant genotypes approaches the value 1.

The process of selection described here could hardly be described as "natural"; we have visualized some external agency which selects the dominant genotypes from the males for breeding. However, the example does give some feeling for the process of evolution by natural selection. In this theory, the selection is supposed to take place because individuals with certain favorable physical characteristics are *more likely* to survive and mate in a competitive environment than those in whom this characteristic is recessive. We have replaced this probabilistic hypothesis (indicated in the "more likely" phrase) by a deterministic selection rule which is much easier to formulate.

Exercises for Section 5.5

1. Prove that the solution of the relations (9), together with the initial conditions (8), is given by Eqs. (10). Use these equations to verify that

$$d_n + 2h_n + r_n = 1 \qquad (n = 1, 2, \ldots).$$

2. Repeat the analysis of Section 5.5 with the exception that the males selected for breeding are all hybrid. Show that, for $n = 0, 1, 2, \ldots,$

$$d_{n+1} = \tfrac{1}{2}d_n + \tfrac{1}{4}(2h_n)$$
$$2h_{n+1} = \tfrac{1}{2}d_n + \tfrac{1}{2}(2h_n) + \tfrac{1}{2}r_n$$
$$r_{n+1} = \phantom{\tfrac{1}{2}d_n +} \tfrac{1}{4}(2h_n) + \tfrac{1}{2}r_n$$

and, hence, that $2h_n = \tfrac{1}{2}$ for $n = 1, 2, \ldots$. It follows that

$$d_{n+1} = \tfrac{1}{2}d_n + \tfrac{1}{8}$$
$$r_{n+1} = \tfrac{1}{2}r_n + \tfrac{1}{8}$$

Prove that the explicit solution is then

$$d_n = \frac{1}{2^{n-1}}(d_0 + h_0) + \frac{1}{4}\left(1 - \frac{1}{2^{n-1}}\right)$$

$$r_n = \frac{1}{2^{n-1}}(h_0 + r_0) + \frac{1}{4}\left(1 - \frac{1}{2^{n-1}}\right)$$

and, hence, that as n increases d_n and r_n each approach the value $\frac{1}{4}$. Are these the results to be expected on intuitive grounds?
3. The models of this section are open to criticism on the grounds that, in many species, dominant and hybrid individuals are indistinguishable. Study the case in which all parents, male and female, are either dominant or hybrid.

5.6   Gene Frequencies

Suppose that a population of N individuals consists of dN dominants, $2hN$ hybrids, and rN recessives. We can easily calculate the total numbers of dominant and recessive genes. Thus, there are $dN + hN$ dominant and $hN + rN$ recessive genes. The set of *all* genes in the population ($2N$ in number) is called the *gene pool*, and the proportions

$$p = d + h, \qquad q = h + r$$

are called the *gene frequencies* of the population. Note that $p + q = 1$.

If the population is propagated under the panmixia model developed in Sections 5.3 and 5.4, then the relations (5) obtain and we can write, for $n = 0, 1, 2, \ldots$,

$$(11) \qquad d_{n+1} = p_n^2, \qquad 2h_{n+1} = 2p_n q_n, \qquad r_{n+1} = q_n^2.$$

It then follows that

$$p_{n+1} = d_{n+1} + h_{n+1} = p_n^2 + p_n q_n = p_n$$
$$q_{n+1} = h_{n+1} + r_{n+1} = p_n q_n + q_n^2 = q_n.$$

Thus we deduce that, *for a population in equilibrium* under the Hardy-Weinberg law, the gene frequencies do not vary from one generation to another. Since we know that d_n, h_n, r_n are fixed for all $n \geq 1$, the only new piece of information here is that there is no change in gene frequencies from the zeroth to the first generation!

Thus, given the gene frequencies p and q of a population, we can immediately deduce (from Eqs. (11)) what the genotype frequencies would be in equilibrium under panmixia

$$d = p^2, \qquad 2h = 2pq, \qquad r = q^2.$$

In contrast to this situation it is clear that, in the model with selection discussed in Section 5.5, the recessive gene frequency q_n decreases to zero as n increases. Consequently, p_n increases simultaneously to 1 (see Exercise 3).

Exercises for Section 5.6

1. Consider a population in equilibrium under panmixia. Sketch graphs of d, $2h$, and r as functions of the recessive gene frequency q.
2. In humans, *albinism* (lack of pigment deposition) is an infrequent phenomenon for which the model of a *recessive* genotype is appropriate. Thus, people who are *not* albino are either dominant or hybrid. In a survey it is found that there is a frequency of albinos of 49 in a million. Assuming an equilibrium population under panmixia, show that the gene frequencies are given by $p = 0.993$, $q = 0.007$. Show that in such cases, where q is very small, the proportions of dominants, hybrids, and recessives are given approximately by $d = 1 - 2q$, $2h = 2q$, $r = 0$.
3. Prove that, in the model of selection in Section 5.5, the recessive gene frequency in generation n is given by

$$q_n = \frac{1}{2^n}(h_0 + r_0) \qquad (n = 0, 1, 2, \ldots),$$

   and hence that $p_n \longrightarrow 1$ as $n \longrightarrow \infty$.
4. In Exercise 9 of Section 5.4, show that the genotype frequencies satisfy the relations

$$p_{n+1} = \frac{1}{1 + q_n}, \qquad q_{n+1} = \frac{q_n}{1 + q_n}.$$

Hence prove that

$$p_n = \frac{1 + (n-1)q_0}{1 + nq_0}, \qquad q_n = \frac{q_0}{1 + nq_0} \qquad (n = 1, 2, 3, \ldots).$$

Examine the behavior of the gene frequencies for large n, and discuss their physical significance.

5.7 A Model of the Mutation Phenomenon

We have observed that the Hardy-Weinberg Law gives a convincing model (under appropriate conditions) for the observed stability of hereditary phenomena. However, changes do occur and they are not all explicable by the "selection" idea introduced in Section 5.5. It is observed experimentally that, on rare occasions (perhaps in one of 10^5 or 10^6 trials), a recessive genotype will occur when, on our modeling to date, it should *not* occur. Such an occurrence is called a *mutation*.

In order to model this phenomenon, we now assume that, whenever a dominant gene is transmitted, there is a probability α $(0 \le \alpha \le 1)$, generally very small, that the gene will *mutate* to a recessive gene. We suppose that this mutation occurs *after* selection of the dominant gene from a parent. Otherwise, we retain all the hypotheses of the panmixia model. What trends in genotype frequencies would we expect? Using the gene frequency notion introduced above, the following is clear: Given a frequency of dominant genes p_n in generation n, then, since $1 - \alpha$ is the probability of *no* mutation, we should now expect

$$(12) \qquad p_{n+1} = (1 - \alpha)p_n.$$

This is in contrast to the Hardy-Weinberg Law which implies $p_{n+1} = p_n$. Since $\alpha > 0$ implies $1 - \alpha < 1$, we see that the dominant gene frequency will decrease steadily, and we therefore anticipate steadily decreasing frequencies of the dominant and hybrid genotypes.

We now compute the genotype frequencies. Let D_m, R_m, D_f, R_f have the same meaning as in the model of section 5.2 *except* that the perturbation due to mutation is to be included. For example, D_m is the event that the gene selected at random from a randomly selected male is dominant *and* does not mutate. If we use M, N to denote occurrence and non-occurrence of mutation of a dominant gene, then we can extend the sample space of Eq. (4) to admit mutation as follows:

$$(13) \qquad \{\delta_m DN, \gamma_m DN, \delta_m DM, \gamma_m DM, \gamma_m R, \rho_m R\}$$

and the first two outcomes give rise to the ultimate transmission of a dominant gene, i.e. event D_m. The last four outcomes represent event R_m. Note that $\delta_m D, \gamma_m D, \gamma_m R, \rho_m R$ have the same probabilities as before: d_n, h_n, h_n and r_n. Since N and M have probabilities $1 - \alpha$ and α respectively, it follows that

$$P(D_m) = P(\delta_m DN) + P(\gamma_m DN) = (d_n + h_n)(1 - \alpha) = (1 - \alpha)p_n.$$
$$P(R_m) = 1 - P(D_m) = (p_n + q_n) - (1 - \alpha)p_n = \alpha p_n + q_n.$$

Exactly the same arguments apply to the genes transmitted by females so $P(D_f) = (1 - \alpha)p_n$ and $P(R_f) = \alpha p_n + q_n$.

Thus, the probabilities that an individual selected at random from the $(n + 1)$th generation is of the dominant, hybrid, or recessive genotypes are

$$
\begin{aligned}
& d_{n+1} = P(D_m)P(D_f) = (1 - \alpha)^2 p_n^2, \\
(14) \quad & 2h_{n+1} = P(D_m)P(R_f) + P(R_m)P(D_f) = 2(1 - \alpha)p_n(\alpha p_n + q_n), \\
& r_{n+1} = P(R_m)P(R_f) = (\alpha p_n + q_n)^2.
\end{aligned}
$$

As a check on our calculations, we observe that

$$p_{n+1} = d_{n+1} + h_{n+1} = (1 - \alpha)p_n\{(1 - \alpha)p_n + (\alpha p_n + q_n)\} = (1 - \alpha)p_n$$

which we obtained in Eq. (12). This recurrence relation yields

$$p_n = (1 - \alpha)^n p_0,$$

and hence

$$\alpha p_n + q_n = \alpha p_n + 1 - p_n = 1 - (1 - \alpha)p_n$$
$$= 1 - (1 - \alpha)^{n+1} p_0.$$

Substituting these expressions in Eqs. (14) and writing $\pi_n = (1 - \alpha)^{n+1} p_0$, we obtain the explicit solution:

$$d_{n+1} = \pi_n^2$$
$$2h_{n+1} = 2\pi_n(1 - \pi_n) \qquad (n = 0, 1, 2, \ldots).$$
$$r_{n+1} = (1 - \pi_n)^2$$

Observe that when $\alpha = 0$ there is no mutation, $\pi_n = p_0$ and $d_{n+1} = p_0^2$, $2h_{n+1} = 2p_0 q_0$, $r_{n+1} = q_0^2$ the fixed equilibrium distribution discussed in the previous section. It is clear that if $\alpha > 0$, then as $n \to \infty$, $\pi_n \to 0$, and hence $d_{n+1} \to 0$, $2h_{n+1} \to 0$, and $r_{n+1} \to 1$. Thus, the effect of mutation is, as expected, to produce a population of recessive type, if this is the only effect which is acting to disturb the equilibrium of the genotype distribution. To see how slowly this takes effect, however, suppose we take for α the value 10^{-5} and seek the effect on the 100th generation. Observe that, using the binomial theorem (Appendix 1 of Chapter 3), we can make the approximation

$$\pi_{99} = (1 - 10^{-5})^{100} p_0 \simeq (0.999) p_0,$$

and the coefficient is accurate to three decimal places. Then

$$1 - \pi_{99} \simeq 1 - (0.999) p_0 = q_0 + (0.001) p_0.$$

The resulting genotype distributions are tabulated in the following table and compared with the corresponding equilibrium distribution (which obtained at generation zero).

	With Mutation	Equilibrium
d_{100}	$(0.998)p_0^2$	p_0^2
$2h_{100}$	$(1.998)p_0 q_0 + (0.002)p_0^2$	$2p_0 q_0$
r_{100}	$q_0^2 + (0.002)p_0$	q_0^2

In most living species, it appears that the selection and mutation processes come into play simultaneously. In this chapter we have, for simplicity, separated the two effects. The reader will have observed that, as we have formulated the problems, the selection effect, if acting in isolation, will tend to produce a population dominated by the dominant gene. In contrast, the mutation effect favors the recessive genes. Thus, our first model which predicted equilibrium is subject to (at least) two opposed effects which may vary in intensity in different environments. We are clearly on the brink of rather more complicated and subtle phenomena which would require more biological information, as well as mathematical expertise, to arrive at satisfactory models.

As a final challenge, the reader may wish to investigate a single model on the lines of this chapter which will include the effects of both selection and mutation.

Exercises for Section 5.7

1. Show that in the model of this section,

$$d_n = p_n^2, \qquad 2h_n = p_n q_n, \qquad r_n = q_n^2, \qquad (n = 1, 2, 3, \ldots)$$

(as in the Hardy-Weinberg equilibrium state).

2. (a) Let G_1 be the event that, in a random choice from the gene pool, a dominant gene is selected. Let G_2 be the event that, in the random selection of a male followed by the random selection of a gene from the male selected, the outcome is a dominant gene. Show that $P(G_1) = P(G_2)$.

   (b) Use part (a) and Eq. (12) to establish Eqs. (14).

Preliminary Exercises

The reader should have some mastery of the concept of recurrence relations and the solution of linear recurrence relations. There is ample material of this kind in Chapter 3. In that chapter, Preliminary Exercises 5, 6 (geometric series), and 14–16 (induction) will be useful. Section 3.10 contains a discussion of three-term recurrence relations in some generality; that discussion is more than we need in this chapter, where ad hoc methods suffice.

1. [Recall that, if A, B are sets, then the *union* of A and B and the *intersection* of A and B are the following sets:

$$A \cup B = \{x \colon x \in A \text{ or } x \in B\},$$
$$A \cap B = \{x \colon x \in A \text{ and } x \in B\},$$

respectively.]

If $A = \{-2, 3, 4\}$ and $B = \{-4, -2, 0, 1, 4, 5\}$, write $A \cup B$ and $A \cap B$. Write a list of all the subsets of A. (A set of n members has 2^n subsets.)

2. With A, B as in Preliminary Exercise 1, let $C = \{-3, 0, 4, 5\}$. Verify that, in this case, $A \cup (B \cup C) = (A \cup B) \cup C$ and $A \cap (B \cap C) = (A \cap B) \cap C$. Then prove these statements for arbitrary sets, A, B, and C.

CHAPTER SIX

COLLISIONS OF PARTICLES

6.1 Introduction

In this chapter we will discuss a class of problems from mechanics. Historically, the study of mechanics has probably been the most important single subject of study in applied mathematics. Indeed, until the end of the nineteenth century, every mathematician worth his salt was familiar with mechanics and, particularly, with the applications to astronomy. Until this time, the development of mathematics itself was stimulated to a great extent by a desire to understand and describe natural phenomena; many of these involved the motion of particles, rigid bodies, and fluids. Newton himself provides a fine example of this. He developed the basic ideas of calculus largely as an aid to understanding and describing problems in mechanics and astronomy.

We shall discuss mathematical models of collision processes. The reason for this choice is very simple: These problems can be studied without the aid of calculus and are nevertheless significant and interesting problems. As we shall see, we can even go as far as twentieth-century developments and introduce the formulation of the problem in the context of the theory of special relativity. It must be understood that, in order to make this subject area accessible, we must confine our attention to the rather artificial case of motion on a straight line. This is a severe limitation which can lead only to a very limited understanding of special relativity. However, we have three reasons for pursuing this subject in spite of the several difficulties. First, there is the historical perspective which the development gives. Second, the development from classical to relativistic theory provides a good example of the successive refinement of mathematical models. Third, the subject has some glamour, and so this early introduction may stimulate the reader to further study in the area.

First we must discuss some of the basic concepts of mechanics. It should be borne in mind that we are entering at once into the business of model-making. The fundamental concepts discussed in the next section form some of the basic ideas from which the models of mechanics are constructed. Notice also that idealization, or abstraction, of the real world is the essential ingredient in this process of forming concepts.

6.2 Some Basic Concepts

We shall introduce only those concepts needed for this chapter. The concept of *force* is notably absent from our list but would have to appear before almost any other problem of mechanics could be discussed. The concepts we introduce form the building blocks for our mathematical models, and the validity of the models produced obviously depends on the validity of these basic concepts. The ultimate test of validity of the models, and hence of the basic concepts, consists of checking behavior predicted by the model against the observed behavior of the physical system.

We will not indulge in a detailed discussion of the nature of space and time, but we will postulate that measurements in space and in time are possible. In order to do this, we first need the abstract concept of a *rigid rod* of invariable length (and, of course, the concept of a *straight line*). We take the length of this rod as a unit of distance for measurements in space. We suppose also that fractions of this unit of length can be marked on the rod and used for measurment. The distance between two points can then be established by laying off the unit rod on the straight line connecting these two points. To measure time, we need the abstract concept of a *clock*. Here we think of some physical experiment which can be repeated indefinitely and, at each repetition, requires precisely the same length of time from beginning to end. This length of time is taken as our unit of time. In a watch, the experiment is the oscillation of a balance wheel. The rotation of the earth with respect to a fixed star may also be chosen as the preferred experiment. In this case, the resulting unit of time is the sidereal day. Neither of these experiments attains the ideal of a precisely reproducible unit of time, but in a great many situations of everyday experience and experimental science, they do provide very satisfactory clocks. It is assumed that, when two identical clocks are brought together, they may be *synchronized*, i.e., they are set so that, as long as they have the same motion, they will give the same reading of time.

We emphasize that, as used subsequently in this chapter, the terms "rigid rod" and "clock" refer to ideas which exist only in our minds. They no longer refer to tangible measuring devices.

A *rigid body* in mechanics is a body that can undergo no changes in size or shape. The relative positions of all points of the body, as measured with our rigid rod, are invariant. Note that we are using the idea of "body" as one of the undefined terms in our development. We shall tacitly admit other undefined terms as we proceed. Another useful concept that depends on the idea of rigidity is a *platform*, or *frame of reference*. The platform may be thought of as a rigid body (which may be considered fixed or in motion) from which measurements in space and time can be made.

It is a fundamental postulate of classical or Newtonian mechanics that identical synchronized clocks will remain synchronized whether (a) at different points fixed in a given platform, or (b) fixed in different platforms which have different velocities. In the theory of special relativity, postulate (a) is accepted, but *not* postulate (b). Our initial approach to collision processes is in the classical vein, but we shall later discuss the interdependence of space and time measurements in order to develop the theory from the point of view of the theory of relativity.

We referred earlier to the possibility of a body being at *rest* or in *motion*. It must be understood that these terms are used in a relative way. "Rest" will always mean fixed with respect to some platform, and we cannot attach any absolute meaning to the phrase "at rest." To illustrate this point, while investigating the motion of a spinning coin (or of a gyroscope) in a spaceship in steady flight near the earth, it may be quite sufficient to describe the motion relative to a platform fixed in the spaceship and to suppose that the platform is at rest. However, to investigate the motion of the spaceship itself relative to the earth, it may be necessary to refer this to a platform fixed in the earth and supposed to be at rest. This in turn may be inappropriate for a detailed description of the path of the spaceship in space, and we may have to make use of a platform fixed in the sun.

In this chapter, we shall always be concerned with the motion of points along straight lines. Since we can measure distance and time, we may also measure the *velocity* of a point moving along a line (relative to a platform). The velocity is simply the rate of change of distance measured along the line with time. Since the motion of the point may be in one of two directions along the line, we always think of velocity as a *signed number*, that is, a positive or negative real number depending on some agreed convention as to the direction of motion determining a positive velocity. When we say that a particle has velocity v, v is a place-holder for any real number: positive, zero, or negative. The addition of velocities is then assumed to be precisely the law of addition for real numbers. We shall use *speed* to denote the absolute value* of velocity, and therefore there is no direction implicit in the use of "speed", though there is in the use of "velocity".

*See Section 3.5 and Preliminary Exercises 3.2 and 3.3 for absolute values.

All concepts introduced to this point are fundamentals of *kinematics*. This is the subject which describes the motion of a body with no reference to the forces or underlying physical laws giving rise to the motion. We now need a rather different attribute of matter known as *mass*. We suppose that a particular piece of matter is selected and is assigned one unit of mass. We suppose further that this body of unit mass is reproducible and can be subdivided to provide fractions of the unit. The masses of other bodies are then obtained by comparison with the unit. One method for doing this is to allow the possibility of a weighing experiment, either on a symmetrical scale or on a spring balance. But either of these seems to depend on the idea of a force of gravity, and since we have not defined force, we are in danger of a circuitous argument. We can avoid this by means of a "collision" experiment. We imagine two bodies moving toward one another along a straight line, each with the same speed. They collide and their speeds after collision are noted (they must be away from one another or at rest, since they cannot pass through each other). If the two speeds after collision are equal, then we postulate that the bodies have the same mass. The mass of a given body can then be found if we find how many units and fractions of units of mass are needed to yield equal speeds after collision with this body. Note that mass is always a positive real number.

In everyday life, a tiny scrap of matter is referred to as a *particle*. The word "tiny" here is used with certain dimensions related to the human body in mind. What is tiny for us may be far from tiny to an ant! A first attempt at a defintion of a particle may be as a piece of matter which may be located at a point in space at a given time, but which has a maximum diameter of zero. But if we complete the abstraction, a particle consists of a certain mass which is associated with a point in space. This point in space may be in motion, and the mass may vary with time or remain constant, depending on the model we need. In practice, this will mean something close to the colloquial usage: a body whose dimensions are very small in comparison with other distances or lengths involved. In discussing the motion of the earth relative to the sun, we may represent the earth as a particle. But the earth cannot be represented as a particle if we discuss the motion of a ball thrown from one child to another.

Exercises for Section 6.2

1. A car drives along a road which is parallel to a railway track. A train traveling at 80 mph passes the car which is traveling in the same direction at 60 mph Measuring velocities in the direction of motion of the train, what is the apparent velocity of the train observed from the car? Of the car observed from the train?

   Answer the same question if the car is traveling at 50 mph in the opposite direction.

Answer the same question if the train has velocity u and the car has velocity v. (Note that the car and the train constitute two *frames of reference*.)

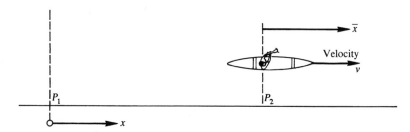

Figure 6.1

2. An observer, P_1, stands on a river bank and measures distance x along the bank (positive in a certain direction). A second observer, P_2, progresses in the positive x direction, with a uniform velocity v, in his canoe (Fig. 6.1). P_2 and P_1 have their watches synchronized, and both agree to measure time from the moment $t = 0$ as P_2 passes P_1. If P_1 and P_2 measure times and distances (t, x) and $(\bar{t}, \bar{x})$, respectively, show that

$$x = \bar{x} + v\bar{t}, \qquad t = \bar{t},$$

or equivalently,

$$\bar{x} = x - vt, \qquad \bar{t} = t.$$

6.3   Perfectly Elastic Collision Processes

We now consider the problem of two particles of different masses which approach each other along a straight line and collide. The particles do not disintegrate, but rebound in some way with no change in mass. The sort of question one might ask is: Given the masses and velocities of the particles just before in.pact, what are their velocities after impact? Note that we refer here to *velocities*. This will allow us to deal simultaneously with collisions which are head-on or which occur as one particle overtakes the other. We should also bear in mind that we are now in a second phase of the model-building process. Our model may represent (with varying accuracy) the collision of cars on a street, of balls on a billiard table, or of atoms in space. Only by checking the results of our subsequent theory with experiment can we decide whether our model will be adequate or not. Notice that the procedure of checking the final model with experiment means that, in cases of failure, the applied mathematician must decide where the fault lies. He may then have to readjust the hypotheses made about the model, or he may even have to reformulate his fundamental concepts.

Let us focus on the case of billiard balls for a moment. If we attempt to study their motion by replacing each ball by a particle, then we are assuming (among other things) that, for the purposes of this study, all effects resulting from the finite size of the balls are negligible. For example, we must ignore the effect of any spin which either ball may have, the roughness of their surfaces and, initially at least, the effect of elastic distortion of the balls on impact. These assumptions are part of the process of *idealization* that goes into making the model. We do not imply that the effects we presently ignore will always be unimportant. When they are significant, our model is likely to be inadequate, and we must modify it or make a fresh start on a more sophisticated model of the situation.

In order to make predictions about the behavior of the particles after collision, we have to invoke two principles of mechanics. We can do little more than quote these principles, but their origin lies in centuries of investigation and thought stimulated by experimental evidence. We use these principles simply because they are found to work in appropriate situations.

We first need the idea of *momentum*. A particle of mass m, having velocity v, is said to have the momentum mv. Note that momentum is also a signed number. A 200-lb ice-hockey player racing over the ice seems much more formidable than a 100-lb child skating at the same velocity. Why? The following argument will show one reason to be that the 200-lb man has twice the momentum of the child.

The first principle is called the *conservation of momentum*. For collision processes, it states that the total momentum of the two particles is the same before and after impact. We can formulate this principle as follows. Let the particles have masses m_1, m_2 and velocities u_1, u_2 before impact and v_1, v_2 after impact, respectively. Note that the masses do not change and are supposed to be nonzero. Then

$$(1) \qquad m_1u_1 + m_2u_2 = m_1v_1 + m_2v_2.$$

In the problem posed above, we are given m_1, m_2, u_1, and u_2 and wish to determine v_1 and v_2. The principle of conservation of momentum yields a linear equation connecting these two unknowns, but it does not allow us to determine either of them.

By rearranging the above formulation of the principle, we obtain

$$m_1(u_1 - v_1) = -m_2(u_2 - v_2)$$

and hence, provided $u_1 \neq v_1$, we have

$$\frac{m_1}{m_2} = -\frac{u_2 - v_2}{u_1 - v_1}.$$

If we interpret $u - v$ as the *change in velocity* due to impact, we can reformulate the principle as follows: The changes in velocity of the two particles are inversely proportional to their masses and are opposite in direction. Thus if m_2 is massive compared to m_1 (the ice-hockey player compared to the boy, for example), then $u_2 - v_2$, the change in velocity of m_2 (the hockey player), is very small compared to that in m_1 (the boy), as our intuition would suggest. If $m_1 = m_2$, the changes in velocity have the same magnitude. (We see at this point that some foreknowledge of the principle of momentum was implicit in our definition of mass.) Note also that, if $u_1 = v_1$, then it follows that $u_2 = v_2$, and the velocities of the two particles are unchanged. This would imply no collision, and so we may safely assume throughout that $u_1 - v_1 \neq 0$ and $u_2 - v_2 \neq 0$.

Newtonian mechanics can be based on three fundamental laws—which do not include the principle of momentum. However, with just a little mathematical machinery, the principle of momentum can be derived from these laws. The next principle we need, an energy law, cannot be derived directly from Newton's laws, but depends on a closer study of the impact process. We will study a case known as *perfectly elastic* impact. This is a highly idealized situation in which it must be assumed, for example, that there is no deformation of the colliding bodies on impact. Rather than attempting to justify this principle from a discussion of the forces of interaction between particles, we again present this as a cut-and-dried formula justifiable by experiment under appropriate conditions.

The *kinetic energy* of a particle with mass m and velocity v (speed $|v|$) is defined to be the nonnegative quantity $\frac{1}{2}mv^2$. The principle of *conservation of energy* (for perfectly elastic impact) says that the total kinetic energy is the same before and after impact. Thus, in the above notation,

$$(2) \qquad \tfrac{1}{2}m_1u_1^2 + \tfrac{1}{2}m_2u_2^2 = \tfrac{1}{2}m_1v_1^2 + \tfrac{1}{2}m_2v_2^2.$$

We now show that these two principles, the conservation of momentum and of energy, are together sufficient to determine v_1 and v_2 if $m_1, m_2, u_1,$ and u_2 are known.

Before plunging into this, let us pause and note that it is now suggested that step 2 of our general procedure (Chapter 1) is complete. We have our mathematical model, and we are now embarking on step 3 which consists of solving Eqs. (1) and (2) for v_1 and v_2. All that is involved here is a little ingenuity in algebraic manipulation.

First, rewrite Eq. (2) in the following way:

$$m_1u_1^2 - m_1v_1^2 = -m_2u_2^2 + m_2v_2^2.$$

This implies

$$m_1(u_1^2 - v_1^2) = -m_2(u_2^2 - v_2^2)$$

or

$$(3) \qquad m_1(u_1 - v_1)(u_1 + v_1) = -m_2(u_2 - v_2)(u_2 + v_2).$$

But we have seen that the momentum law (1) gives

$$m_1(u_1 - v_1) = -m_2(u_2 - v_2).$$

If we make this substitution in the left side of Eq. (3) and note that $m_2 \neq 0$, $u_2 - v_2 \neq 0$, we obtain

$$u_1 + v_1 = u_2 + v_2$$

or the so-called *velocity relation*:

(4) $$u_1 - u_2 = -(v_1 - v_2).$$

In other words, the velocity of approach before impact, $u_1 - u_2$, is the negative of the velocity of separation after impact, $v_1 - v_2$. To see the physical significance of this, imagine a platform fixed in the second particle. From this platform the first particle is observed to approach with velocity $u_1 - u_2$ and to have a velocity $v_1 - v_2$ after impact. The velocity relation tells us that these quantities have the same magnitude but opposite sign. The first particle would be observed to "bounce off" the second and recede with the same speed as that with which it collided. This is why this model is described as that of "perfectly elastic" collision.

To solve the problem posed initially, it is now a straightforward calculation to solve the simultaneous equations obtained from (1) and the velocity relation for v_1 and v_2. It is found that

(5)
$$v_1 = \frac{(m_1 - m_2)u_1 + 2m_2u_2}{m_1 + m_2}$$

$$v_2 = \frac{2m_1u_1 - (m_1 - m_2)u_2}{m_1 + m_2}$$

It is often the case in mathematics that examination of extreme or special cases is illuminating, and this is no exception. Suppose we examine the case in which m_2 is very massive compared to m_1 and in which this massive particle is initially at rest, i.e., $u_2 = 0$. Then in this special case, Eqs. (5) yield

$$v_1 = \left(\frac{\mu - 1}{\mu + 1}\right)u_1 \quad \text{and} \quad v_2 = \frac{2\mu}{\mu + 1}u_1$$

where we have written $\mu = m_1/m_2$. Now our hypothesis means that μ is very small compared to 1. We shall go to an extreme case now and suppose that, for all practical purposes, we may set $\mu = 0$. This would be a reasonable assumption if m_2 represented the mass of the earth and m_1 represented the mass of a ball bouncing on the ground. Then we find that v_1, the velocity of rebound of the ball, is just $-u_1$, the negative of the velocity with which the ball hits the ground. This is what we would expect of a perfectly elastic ball. We also find that $v_2 = 0$. The velocity imparted to the earth is zero, which we would also expect.

This example also shows that our model is likely to be inadequate for many problems. By observation of the steadily decreasing height to which a ball bounces, we know that the speed with which a ball rebounds is generally less than that with which it strikes the ground. A model that accounts for this phenomenon is discussed in Exercise 3 of Section 6.5.

Exercises for Section 6.3

1. Complete the derivation of Eqs. (5) from Eqs. (1) and (4).
2. A particle of mass 2 tons with velocity 60 mph collides with a particle of mass $\frac{1}{2}$ ton and velocity 30 mph. What are the velocities of the particles after collision?
3. Show that, in a perfectly elastic collision on a straight line, particles of equal mass exchange velocities.
4. Two elastic balls of masses 1 and 2 and velocities 2 and 1, respectively, are in collision, and the leading ball then hits a third elastic ball of mass 2 and velocity $\frac{2}{3}$. Will there be any further collisions?
5. If particles of mass 100 grams and 200 grams are observed to have velocities $v_1 = -20$ cm/sec and $v_2 = 40$ cm/sec, respectively, *after* a perfectly elastic impact, what were their velocities before impact?
6. In the perfectly elastic collision of particles on a straight line, let p_T, e_T be the total momentum and kinetic energy of the two particles, respectively. (Thus, the momentum law gives $m_1 u_1 + m_2 u_2 = p_T$ and $m_1 v_1 + m_2 v_2 = p_T$, for example.) Then the momenta before and after impact are solutions of the momentum equation,

$$\pi_1 + \pi_2 = p_T,$$

and also the energy equation,

$$\frac{\pi_1^2}{m_1} + \frac{\pi_2^2}{m_2} = 2e_T.$$

Sketch graphs of these two equations using π_1, π_2 coordinate axes in the following cases:
(a) $m_1 = m_2 = 1$; $u_1 = 4$, $u_2 = -3$.
(b) $m_1 = 1$, $m_2 = 4$; $u_1 = 4$, $u_2 = -3$.
Use the graphs to estimate v_1, v_2 in both cases, and compare your estimate with the exact solution.
7. Two particles have masses 3 and 4 and a total energy of 6. What is the maximum (a) momentum and (b) relative speed ($|u - v|$) they can have?

6.4 Dependence on the Platform

We remarked earlier that we cannot establish an absolute measure of velocity. We now wish to examine our model in the light of this remark. In our earlier discussion, a certain platform of observation P_1 was implicit in the statements "let the velocities before impact be u_1, u_2," and so on. Suppose the measurements of velocity were made from a second platform, P_2, moving with velocity u as measured from P_1. Observed from P_2, the velocities of the particles in the collision process would all seem to be reduced by u. Thus, the velocities before and after impact would be $u_1 - u, u_2 - u$ and $v_1 - u$, $v_2 - u$ respectively.

To illustrate this apparent change in velocity, suppose that a man standing on the ground sees a train go by at 60 mph. He makes this observation from a platform P_1 fixed on the ground. On the other hand, to an observer in a car traveling parallel to the train and in the same direction at 40 mph, the train has an apparent velocity of 60 mph − 40 mph = 20 mph. This is observed from his platform P_2 fixed in his car. The reader should check the case in which the car travels in the opposite direction.

To return, now, to the general discussion, do the conservation laws assumed for P_1 imply corresponding laws for the same collision phenomenon as observed from P_2? The answer is yes. Let us repeat what we are trying to do. We assume the momentum and energy laws, formulated in Eqs. (1) and (2), to hold for all collisions observed from P_1. We also assume that velocities measured from platform P_2 will be those observed from P_1 reduced by u. We want to know whether these assumptions imply that an observer in P_2 can use momentum and energy laws, but formulated in terms of the velocities as he measures them.

To see this for the momentum law, we simply subtract $m_1u + m_2u$ from both sides of Eq. (1) to obtain the appropriate momentum equation for P_2:

$$m_1(u_1 - u) + m_2(u_2 - u) = m_1(v_1 - u) + m_2(v_2 - u).$$

To obtain the energy law for platform P_2, we first observe that Eq. (1) (a momentum equation!) implies

$$m_1u_1u + m_2u_2u = m_1v_1u + m_2v_2u.$$

We subtract corresponding sides from Eq. (2) and add $\frac{1}{2}m_1u_1^2 + \frac{1}{2}m_2u_2^2$ to both sides. Then we deduce that

$$\frac{1}{2}m_1(u_1 - u)^2 + \frac{1}{2}m_2(u_2 - u)^2 = \frac{1}{2}m_1(v_1 - v)^2 + \frac{1}{2}m_2(v_2 - v)^2,$$

which is the energy law for observations from P_2.

It is interesting that, in order to show that the energy law is independent of the observation platform, we had to use both the momentum and the energy laws. This suggests some intrinsic relationship between the two laws; this observation will be of interest to us when we consider impacts in relativity theory.

The lack of dependence on the platform of observation is a property of the model we may have hoped for, and the property is reassuring. We should bear in mind, however, that although our confidence in the model is increased as a result of this property, the justification of the model still depends primarily on experimental verification.

Exercises for Section 6.4

1. Solve the problem of Exercise 2, Section 6.3, by considering the velocities as observed from a platform moving with the smaller particle before collision and then reverting to the platform from which the velocities are given. Solve the problem again by using velocities measured from a platform moving with the larger particle.

2. If masses m_1, m_2 are at points P, Q, respectively, the *center of mass* of m_1 and m_2 is the point R on the segment PQ such that

$$\frac{|PR|}{|RQ|} = \frac{m_2}{m_1}.$$

If m_1, m_2 have velocities u_1, u_2, show that R has velocity

$$\frac{m_1 u_1 + m_2 u_2}{m_1 + m_2}.$$

Hence show that the velocity of the center of mass is not changed in a perfectly elastic collision.

3. Show that, in Exercise 2, the *speed* of m_1 relative to the center of mass is unchanged in the collision.

6.5 Inelastic Impact

There are many collisions for which our previous model will not be appropriate. One obvious case is a collision in which the particles are known to stick together after impact. For example, a collision between railway trucks which are automatically coupled together on impact. We shall describe such a collision as *perfectly inelastic*.

Again, it has been observed experimentally that momentum is conserved in such processes; so we take this as an axiom for our model. If v is the velocity of the composite particle after impact and $m_0 = m_1 + m_2$, then the momentum after impact is $m_0 v$, and the law of conservation of momentum now reads

$$m_1 u_1 + m_2 u_2 = m_0 v.$$

If we ask again for the velocity after impact given the initial velocities, we see that this law alone is sufficient to solve the problem:

(6) $$v = \frac{m_1 u_1 + m_2 u_2}{m_0}.$$

The mathematical model is more simple in this case, since we do not need to invoke an energy law. However, let us give rein to our scientific curiosity and compare the total kinetic energies before and after impact. Before impact we have $\frac{1}{2}m_1 u_1^2 + \frac{1}{2}m_2 u_2^2$, as before, and we wish to compare this with $\frac{1}{2}m_0 v^2$. Using Eq. (6), we have

$$\frac{1}{2}m_0 v^2 = \frac{1}{2m_0}(m_1 u_1 + m_2 u_2)^2$$

$$= \frac{1}{2m_0}\{m_1^2 u_1^2 + m_2^2 u_2^2 + 2m_1 m_2 u_1 u_2\}$$

$$= \frac{1}{2m_0}\{(m_1^2 u_1^2 + m_1 m_2 u_1^2) + (m_2^2 u_2^2 + m_1 m_2 u_2^2)$$

$$+ (2m_1 m_2 u_1 u_2 - m_1 m_2 u_1^2 - m_1 m_2 u_2^2)\}$$

$$= \frac{1}{2m_0}\{m_0 m_1 u_1^2 + m_0 m_2 u_2^2 - m_1 m_2 (u_1 - u_2)^2\}$$

$$= \frac{1}{2}m_1 u_1^2 + \frac{1}{2}m_2 u_2^2 - \frac{1}{2}\frac{m_1 m_2}{m_0}(u_1 - u_2)^2.$$

Thus, the final kinetic energy is equal to the initial kinetic energy minus the quantity

$$g = \frac{1}{2}\frac{m_1 m_2}{m_0}(u_1 - u_2)^2$$

which we may refer to as the kinetic energy lost in the impact.

Now it can be argued that the energy g is not "lost", or dissipated, but merely used up in some other way which cannot be discovered if we confine our attention to the highly idealized model. For example, in the collision of material bodies, there is generally deformation of the bodies themselves. A more complicated model could perhaps be set up which would allow us to account for deformations of the bodies as well as their velocities before and after impact. In such a case, we would expect to retain the momentum law, and hence the result (6), but we would also be able to predict other phenomena in the collision and, in particular, we would be able to interpret g as an energy, perhaps of a different form. As long as we stay with our particle model, we shall refer to g as an *internal energy* lost in the impact.

This allows us to retain some form of energy law:

$$\tfrac{1}{2}m_1u_1^2 + \tfrac{1}{2}m_2u_2^2 = \tfrac{1}{2}m_0v^2 + g$$

and although it adds nothing to our solution (6), it will help in the discussion of the next problem.

Observe carefully that the energy equation we have derived is a *deduction* from our model and is not one of the hypotheses which constitute the model. As indicated in Appendix 1, this model consists of the basic concepts discussed in Section 2.2 together with the hypotheses:

1. Two particles combine in the collision to a single particle.
2. The total mass remains constant.
3. The total momentum is conserved.

It is easily seen that these hypotheses are again independent of the platform from which the collision is observed. Incidentally, this observation can be very useful in calculations (Exercise 3(c), for example) and in discussing qualitative results. Thus, there is often a preferred frame of reference from which the mathematical description of a phenomenon takes on a particularly simple form. For example, in the above problem we may choose as our platform P_2 one with the velocity v of the resulting particle. Thus, from P_2 the combined particle has zero velocity, and the velocities before collision are $u_1 - v$ and $u_2 - v$. The momentum equation is then

$$m_1(u_1 - v) + m_2(u_2 - v) = 0,$$

and the energy lost in the impact is

$$g = \frac{1}{2}\frac{m_1m_2}{m_0}(u_1 - u_2)^2 = \frac{1}{2}m_1(u_1 - v)^2 + \frac{1}{2}m_2(u_2 - v)^2.$$

We shall see this device used again in Section 6.10.

Exercises for Section 6.5

1. Particles of mass 3 grams and 5 grams have speeds 20 m/sec and 30 m/sec, respectively, in opposite directions just before colliding. Calculate their velocities after impact if (a) the collision is perfectly elastic, and (b) the collision is inelastic.
2. Particles of mass 10 lb and 50 lb are in a perfectly inelastic collision. If the 10-lb particle had an initial speed of 30 ft/sec. and if the final (combined) particle has a speed of 150 ft/sec. in the opposite direction, what was the speed of the 50-lb mass just before collision?

3. The velocity relations for perfectly elastic and for inelastic impact are

$$v_1 - v_2 = -(u_1 - u_2) \quad \text{and} \quad v_1 - v_2 = 0.$$

There are many collision processes with "intermediate" bounce properties which may be described by a law of the form

(E1) $$v_1 - v_2 = -e(u_1 - u_2), \quad (0 < e < 1).$$

The *coefficient of restitution*, e, determines the degree of elasticity (roughly speaking). The case $e = 0$ is inelastic and $e = 1$ is perfectly elastic. Taking the momentum law and (E1) as the fundamental equations for a general collision process, examine:
(a) The formulas for v_1, v_2 in terms of m_1, m_2, u_1, u_2, and e.
(b) The question of dependence on the platform.
(c) The energy lost in a collision.
4. Show that in a perfectly elastic (Newtonian) collision between particles of equal mass, the velocities of the particles are interchanged (Exercise 3 of Section 6.3).
   A ball of mass m is moving with a velocity u along a straight line towards a row of $n - 1$ balls all with the same mass, m, which are at rest on the line at equal distances apart. If all of the balls are perfectly elastic, what are the final velocities of the balls?
   If the $n - 1$ balls on the line are made of clay so that all of the collisions are perfectly inelastic, what is the final velocity of the composite particle?

6.6 Explosions

We now try to visualize an inelastic impact in reverse! Two bodies (particles) initially joined together are suddenly separated and have different velocities. A bullet shot from a gun is a good example. The bullet and gun are initially joined together, and after the explosion the bullet and gun have different velocities—the gun itself has a velocity giving rise to the *recoil* phenomenon. If you can imagine jumping forward from a piece of wood which is standing on very smooth ice, a similar phenomenon arises.
   We again invoke the law of conservation of momentum. If m_0 is the combined mass with initial velocity u, and if m_1 and m_2 are the component masses with velocities v_1 and v_2 after the explosion, respectively, then

(7) $$m_0 u = m_1 v_1 + m_2 v_2.$$

Although this is a "reversed" inelastic collision, the initial data m_1, m_2, and u are no longer sufficient to allow us to calculate the resulting velocities v_1 and v_2. Once again, we have to invoke a second law in order to complete our model, and we turn first of all to the possibility of an energy law. Our study of elastic and inelastic impacts suggests that an energy relation based on the latter case will most likely be appropriate. Thus we will introduce an internal energy concept again, but in this case, it must be an energy initially stored in the composite body and then used up in the process of separation. We therefore postulate a total energy before the explosion of the form $\frac{1}{2}m_0u^2 + g$, where g represents the internal energy which can be made quantitative only on closer investigation of the mechanism of the explosion. The law of conservation of energy now takes the form

(8) $$\tfrac{1}{2}m_0u^2 + g = \tfrac{1}{2}m_1v_1^2 + \tfrac{1}{2}m_2v_2^2,$$

and we trust that we now have a complete model in the sense that we can determine v_1 and v_2 from Eqs. (7) and (8).

Our method of solution is again indirect. We can deduce from Eqs. (8) and (7), with essentially the same analysis as in Section 6.5, that the internal energy is given by

$$g = \frac{1}{2}\frac{m_1m_2}{m_0}(v_1 - v_2)^2,$$

and this implies that

$$v_1 - v_2 = \pm\sqrt{\frac{2m_0g}{m_1m_2}}.$$

We appeal to the physics of the problem to decide which sign is appropriate. Suppose that the particle of mass m_1 is to the left of the other particle after the explosion and we measure velocities positive when the motion is from left to right. Then, since the particles separate, we must have $v_2 > v_1$. With this agreed labeling of the particles, we have

(9) $$v_2 - v_1 = \sqrt{\frac{2m_0g}{m_1m_2}}.$$

We can now combine this with Eq. (7) to obtain the explicit solution

(10)
$$v_1 = u - \sqrt{\frac{2m_2g}{m_0m_1}}$$

$$v_2 = u + \sqrt{\frac{2m_1g}{m_0m_2}}.$$

Let us use this model to investigate the recoil of a gun on firing a relatively light bullet. We suppose the bullet, of mass m_2, and the gun, of mass m_1, to be initially at rest. Thus $u = 0$ in Eqs. (10). The energy, g, is that released by the cartridge when the bullet is fired. We suppose that m_2 is very small compared to m_1, that is, m_2/m_1 is small compared to the number 1. In this case

$$\frac{m_1}{m_0} = \frac{m_1}{m_1 + m_2} \simeq 1, \qquad \frac{m_2}{m_0} = \frac{m_2}{m_1 + m_2} \simeq \frac{m_2}{m_1}$$

and from Eqs. (10) we deduce that

$$v_1 \simeq -\sqrt{\frac{m_2}{m_1}}\sqrt{\frac{2g}{m_1}}, \qquad v_2 \simeq \sqrt{\frac{2g}{m_2}}.$$

Thus v_1, the velocity with which the gun itself recoils, is small since m_2/m_1 is small. The dominant factors in determining the initial velocity of the bullet are, as one might expect, the internal energy, g, stored in the cartridge and the mass of the bullet itself.

We have now discussed three different models in the framework of classical mechanics. The second half of the chapter is devoted to a reexamination of the basic concepts and then an investigation of the same three phenomena in the light of the theory of special relativity. The reader who has not thoroughly mastered the material up to this point, or who prefers not to delve into the intricacies of relativity, should abandon the chapter at this point.

Exercises for Section 6.6

1. Complete the derivation of Eq. (10) from Eqs. (7) and (9).
2. A bullet of mass 100 grams is fired from a gun of mass 5000 grams with an explosive charge giving an energy of $g = 5 \times 10^6$ gram cm^2/sec^2. If the gun is at rest before firing, find the initial velocity of the bullet.
3. A bullet is fired from a gun which is initially at rest. If m_1, m_2 are the masses of bullet and gun, respectively, show that the respective velocities after firing, v_1 and v_2, satisfy $v_1/v_2 = -m_2/m_1$ and $v_1 v_2 = -2g/(m_1 + m_2)$.

6.7 Vectors

Before going on to the next models for collision processes, it is convenient to briefly introduce a new mathematical tool. We define the concept of *vector* here in a very simple and limited way which is, however, adequate for our immediate needs. We define a vector to be an ordered pair of real numbers, and we denote a vector by (a, b) where a, b are two real numbers. It is helpful to remember the geometrical fact that such a vector can be used to determine a point in a plane with respect to Cartesian axes in the plane. There is then a one-to-one correspondence between the set of all vectors and the set of all points in a plane.

If (a, b) and (c, d) are any two vectors, we define their *sum* by

$$(a, b) + (c, d) = (a + c, b + d)$$

where the sums on the right involve the usual addition of real numbers. The geometrical version of this axiom is the *parallelogram law* for vector addition. Thus, let (a, b) and (c, d) correspond to the points A, B in a plane and let $(0, 0)$ correspond to the point O (as in Fig. 6.2(a)). Then it is a simple exercise in coordinate geometry to see that $(a + c, b + d)$ corresponds to the point D obtained by completing the parallelogram $AOBD$.

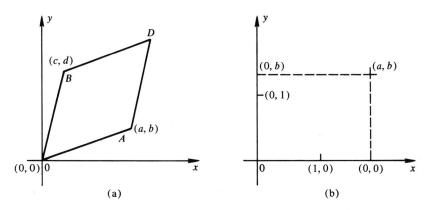

Figure 6.2 (a) The parallelogram law. (b) Representation by unit vectors.

We deduce from the above axiom that

$$(a, b) + (a, b) = (2a, 2b),$$

for example, and we are led to the following definition of *multiplication of vectors by real numbers*:

$$c(a, b) = (ca, cb)$$

for all vectors (a, b) and all real numbers c. These two operations are known as *vector addition* and *scalar multiplication*.

The two vectors $(1, 0)$, $(0, 1)$ play a rather special role and are called the *unit vectors*. We easily verify from the above definitions (cf. Fig. 6.2(b)) that any vector can be expressed in terms of the unit vectors in the following way:

$$(a, b) = a(1, 0) + b(0, 1).$$

6.8   Measurements of Space and Time

The approach we have adopted up to now with respect to measurements of space and time have been accepted and utilized to great advantage for centuries. Indeed, this approach seemed so obvious that, until about a hundred years ago, it was hardly necessary to discuss it explicitly. What was it that persuaded scientists to examine these roots of mechanics? For what physical situations did the old classical mechanics appear to give the wrong results? In other words, where did the mathematical models provided by classical mechanics become inadequate?

This is, of course, a long and involved story. We can only give an outline here and, inevitably, oversimplify it in the process. During the nineteenth century, a new branch of physics was developed which is known as field theory. Faraday, Maxwell, and Hertz figured prominently in this development, and their immediate interest was in the propagation of electromagnetic phenomena: radio waves and light, in particular. Their theory and experiments led to a beautiful mathematical model involving the famous "Maxwell's equations." This theory led to the surprising conclusion that the velocity of light (in a vacuum) is an absolute constant and is the same when measured from any platform of observation. This is clearly at odds with the classical conception of velocity, space, and time, which would allow us to observe light from a platform traveling at the same velocity, thus reducing the apparent velocity of the light to zero.

The only way to resolve the dilemma was to perform experiments involving the velocity of light and measurements of time from platforms having different velocities.

The crucial experimental evidence was provided by physicists Michelson and Morley in 1887. The idea behind the experiment (Fig. 6.3) is very simple. Two rods of the same length, *a*, are joined at right angles and light sent out

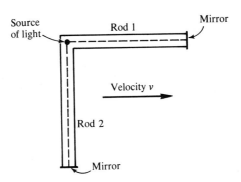

Figure 6.3 The Michelson-Morley experiment.

from the right angle is reflected from mirrors at the ends of the rods. The rods are transported with a known velocity, v, in the direction of one of them. Two hypotheses are made:

1. The speed of light, c, is the same for both beams (as predicted by the electromagnetic theory).
2. The axioms of classical mechanics apply in that the rods can be said to have the same length, a, regardless of their velocities.

On these bases, the time taken for the light to traverse rod 1 and return to the source is found to be $\tau_1 = 2ac/(c^2 - v^2)$. The time taken to return to the source from the mirror on rod 2 is calculated* to be $\tau_2 = 2a/(c^2 - v^2)^{1/2}$. In practice, the velocity, v, that can be arranged for the experiment is *very* small compared to c. Indeed, the common choice for v is the velocity of transport of a point on the surface of the earth relative to the sun, in which case v/c is on the order of 10^{-4}. Thus we have†

$$\tau_1 = \frac{2a}{c}\left(1 - \frac{v^2}{c^2}\right)^{-1} \simeq \frac{2a}{c}\left(1 + \frac{v^2}{c^2}\right)$$

$$\tau_2 = \frac{2a}{c}\left(1 - \frac{v^2}{c_2}\right)^{-1/2} \simeq \frac{2a}{c}\left(1 + \frac{1}{2}\frac{v^2}{c^2}\right)$$

and $\tau_1 - \tau_2 \simeq av^2/c^3$. Although this is likely to be an extremely small difference, the apparatus of Michelson and Morley would be able to detect it. But there was no difference! The experiment has been refined and repeated several times, and these repetitions have only confirmed the original result.

The experimental evidence indicates that at least one of the assumptions (1 or 2) must be false. Einstein decided that assumption 2 was the most suspect and formulated a new kinematics which was consistent with the Michelson-Morely experimental evidence and also formed the basis of the *theory of relativity*. In order to correct the above calculations, we can no longer assume that, even though the two rods have length, a, when they are brought together, they will have the same length when there is a velocity of transport, v, along the length of one of them, but not along the other.

In the sequel, we shall postulate an axiom which ties together measurements in space and time in a rather subtle way. This axiom will then lead to the "relativistic shrinking" of distance measurements which, in turn, yields a satisfactory explanation of the Michelson-Morley result (see Exercise 5 of Section 6.8 and Exercises 6, 7 of Section 6.9).

---

*Note that, as the light approaches the mirror on rod 1, the mirror recedes with speed $|v|$. Pythagoras' theorem is needed for the calculation of τ_2.
†See Appendix 1 of Chapter 3.

For the time being then, rather than discussing the question of measurements of distance, we focus on the measurement of time. The axiom we have referred to will then allow us to turn this discussion to the analysis of the Michelson-Morley experiment.

In the classical theory, the idea of simultaneous events at different locations in space presents no problem. Clocks can be synchronized together, and if one is then moved about in any way relative to the other, the time measurements are always the same. So we can talk sensibly about the same instant of time as measured on either clock. If, however, the time measurement is influenced by the motion of the clock, it is difficult to see what is meant by simultaneous events at different points of a given platform of observation. This is where hypothesis 1 of the Michelson-Morley experiment, the invariance of the speed of light, comes into its own, for we can use this cornerstone of the theory to give a criterion for simultaneous events.

Let A, B be two points and let C be the midpoint of the line segment having A, B as end points. Since the speed of light is an absolute constant, we can say that, whatever the motion of the frame of reference may be, if light signals are emitted from A and B towards each other, then they were emitted *simultaneously* if and only if they both arrive at C at the same instant. This criterion for simultaneity now allows us to talk about *the* time in a given platform. Clocks can be synchronized at different points in space and, provided they subsequently have the same motion, they will remain synchronized.

It is very important to realize that the introduction of the theory of relativity did not mean that the old classical mechanics was no longer valid. Validity of mathematical models depends on reasonable agreement with experiment, and the models of classical mechanics are still the most appropriate in a great variety of physical situations and, in particular, where the speeds in question are all small compared to the speed of light. Indeed, the theory of relativity should reduce to the theory of classical mechanics in such cases, and this is precisely what happens.

We will introduce Einstein's kinematics in the simplest possible case—the description of motion in a straight line—and we will do this in a fairly concise, axiomatic way. Realize that this does *not* correspond to the historical development of these concepts, which would depend on the analysis of experiments in gravitation and electromagnetism which we are not in a position to discuss.

Since space and time measurements are so intimately connected in relativity theory, it is customary to refer to a particular time at a particular point (on a line for our purposes) as an *event*. Thus, to each event corresponds a vector (t, x), where t is the time and x is the position on the line referred to some origin O fixed in a platform of observation. However, it is also customary to use (ct, x), where c denotes the speed of light, rather than (t, x), so that both components of the vector have the dimension of distance. We shall conform with this practice.

Notice that we can now describe *event vectors* in terms of unit event vectors $(1, 0)$, $(0, 1)$ by writing (as in Section 6.7)

$$(ct, x) = ct(1, 0) + x(0, 1).$$

These unit vectors are determined by a clock and a rigid rod which are at rest in the platform of observation. The vector $(0, 0)$ is described as the *origin event*. It corresponds to the origin of the distance coordinate at the initial instant of time.

We must now intrdouce two platforms, P_1 and P_2 (each with its own standard clock and rigid rod), and compare event vectors describing the same point in space and time from the two platforms. Suppose that P_2 has velocity v relative to P_1. In order to distinguish event vectors corresponding to the two platforms, we introduce a subscript notation: $(ct, x)_j$ denotes an event vector measured in platform P_j, for $j = 1$ or 2. For simplicity, we suppose that the two origin events coincide. One more notational convention: If an event vector $(ct, x)_1$ describes the same point in space and time as an event vector $(c\bar{t}, \bar{x})_2$, then we write

$$(ct, x)_1 \longleftrightarrow (c\bar{t}, \bar{x})_2.$$

This does *not* imply equality of the two vectors. However, our assumption that origin events coincide may now be written

$$(0, 0)_1 \longleftrightarrow (0, 0)_2.$$

Consider now any two events described from the two platforms:

$$\text{Event 1:} \quad (ct_1, x_1)_1 \longleftrightarrow (c\bar{t}_1, \bar{x}_1)_2$$
$$\text{Event 2:} \quad (ct_2, x_2)_1 \longleftrightarrow (c\bar{t}_2, \bar{x}_2)_2.$$

Einstein takes as the basic law for his kinematics:

$$(11) \qquad\qquad -c^2 t_1 t_2 + x_1 x_2 = -c^2 \bar{t}_1 \bar{t}_2 + \bar{x}_1 \bar{x}_2.$$

Thus, we postulate a numerical link between observations of any two events from two platforms. We must once more accept this as an axiom without the physical and mathematical insight which led to its first formulation. The reader is asked to take this as an act of faith, or at least as an interesting hypothesis, and to stay with us and see to what conclusions it will lead. Our justification lies in the fact that these conclusions are now firmly established by experiment.

Since our axiom is supposed true for every pair of events $(ct_1, x_1)_1$ and $(ct_2, x_2)_1$, it is true for those cases in which t_1, t_2 have the common value t and x_1, x_2 have the common value x. Thus, for any event $(ct, x)_1 \leftrightarrow (c\bar{t}, \bar{x})_2$,

$$(12) \qquad -c^2 t^2 + x^2 = -c^2 \bar{t}^2 + \bar{x}^2.$$

This is a consequence of Eq. (11) which will be particularly useful.

Let us express Eqs. (11) and (12) in another way. If V_1, V_2 denote the vectors (ct_1, x_1) and (ct_2, x_2), respectively, then we assert that the value of the (bilinear) function, f, defined by

$$f(V_1, V_2) = -c^2 t_1 t_2 + x_1 x_2$$

depends only on the events considered and not on the frames of reference from which observations of the events are made. In other words, f does not depend on the coordinate system chosen. The same remarks apply to the (quadratic) function, g, defined by

$$g(V) = -c^2 t^2 + x^2$$

derived from Eq. (12). The corresponding condition in classical theory is that, when origin events coincide, so do the time observations of any given event, i.e., t, and hence $-c^2 t^2$, is the same measured from any such platform.

Before proceeding to the analysis of the implications of Einstein's law, we must note two important facts which this law and the properties of vector addition and scalar multiplication imply.

Proposition. Let P_1, P_2 be platforms. Let P_2 have a constant velocity as observed from P_1, and suppose that the origin events coincide. We have:

1. If $(ct, x)_1 \leftrightarrow (c\bar{t}, \bar{x})_2$ and k is any real number, then

$$k(ct, x)_1 \leftrightarrow k(c\bar{t}, \bar{x})_2$$

2. If $(ct_1, x_1)_1 \leftrightarrow (c\bar{t}_1, \bar{x}_1)_2$ and $(ct_2, x_2)_1 \leftrightarrow (c\bar{t}_2, \bar{x}_2)_2$, then

$$(ct_1, x_1)_1 + (ct_2, x_2)_1 \leftrightarrow (c\bar{t}_1, \bar{x}_1)_2 + (c\bar{t}_2, \bar{x}_2)_2.$$

This proposition needs proof, of course, but in order to avoid delay in developing our main line of argument, we defer discussion of the proof to Appendix 2 of this chapter.

Exercises for Section 6.8

1. Which of the following corresponding event vectors satisfy Eq. (12):

$$(-2, 4)_1 \longleftrightarrow (2\sqrt{6}, 6)_2, \qquad (4, 2\sqrt{2})_1 \longleftrightarrow (3, 1)_2$$
$$(2 + \sqrt{2}, 2 + \sqrt{2})_1 \longleftrightarrow (2, 2)_2 ?$$

Do any pairs satisfy Eq. (11)?

2. Show that, if $(ct, x)_1 \longleftrightarrow (c\bar{t}, \bar{x})_2$ as in Eq. (12), then

$$\bar{t}^2 - t^2 = \frac{\bar{x}^2 - x^2}{c^2}.$$

If $t = 1$ sec and $\bar{t} = 1.001$ sec, estimate $x - \bar{x}$ if it is given that $c = 186,000$ mi/sec and (a) $x + \bar{x} = 100$ mi, (b) $x + \bar{x} = 10,000$ mi. (This gives some notion of the large scale we must work on to obtain results appreciably different from everyday experience.)

3. You should find in Exercise 1 that the last two pairs of event vectors satisfy both Eqs. (11) and (12). In this case, find the event vectors in P_2 describing the events whose vectors in P_1 are

$$(4\sqrt{2}, 4)_1, \qquad (6 - \sqrt{2}, 3\sqrt{2} - 2)_1, \quad \text{and} \quad (2\sqrt{2}, 4)_1.$$

4. Consider an experiment such as that of Michelson and Morley, except that it is conducted (in still air) with a source of *sound* rather than light. If the velocity of sound in air is ω, what time difference between returning signals is to be expected if they are emitted simultaneously?

5. In this exercise, we show that hypothesis 1 of the Michelson-Morley experiment and the conclusion of that experiment imply the relativistic shrinking of rod 1 in Fig. 6.3. First show that the time taken for the light emitted along rod 1 to return to the source is

$$t_1 = \frac{2a'c}{c^2 - v^2}$$

where a' is the length of rod 1 as measured by an observer moving with it. Then show that the time taken for the beam transmitted along rod 2 to return to the source is

$$t_2 = \frac{2a}{\sqrt{c^2 - v^2}}$$

(use the hints at the foot of p. 144). The conclusion of the experiment is that $t_1 = t_2$. Show that this implies $a' = a\sqrt{1 - v^2/c^2}$ and that $0 < v < c$ then implies $a' < a$, i.e., rod 1 appears shrunk when compared to its length when at rest.

6.9 *The Lorentz Transformation*

Our next objective is to obtain explicitly the relationship between space and time measurements from different platforms. Thus, if $(ct, x)_1 \leftrightarrow (c\bar{t}, \bar{x})_2$, we are to obtain formulas giving t and x explicitly in terms of $\bar{t}$ and $\bar{x}$.

Suppose, first, that we take measurements as observers in P_1. In particular, consider the event corresponding to the space origin of P_2 after time t (as measured in P_1). From P_1, the origin of P_2 is observed to have traveled a distance vt. Thus, the appropriate event vector is $(ct, vt)_1$. Consider the same event viewed from P_2. The space coordinate is zero, and the time measurement is supposed to be τ. Then $(ct, vt)_1 \leftrightarrow (c\tau, 0)_2$ and the law (12) implies that $-c^2t^2 + v^2t^2 = -c^2\tau^2$ or $(c^2 - v^2)t^2 = c^2\tau^2$. This implies that we may take $t = \beta\tau$, where we define

$$(13) \qquad \beta = \frac{1}{\sqrt{1 - v^2/c^2}}.$$

Furthermore, we deduce that for no velocity, v, can $|v|$ reach or exceed the speed of light, c. Thus the denominator in Eq. (13) is always positive and β is well-defined.

We see at once that, since $t = \beta\tau$, the times associated with the same event in two different platforms can differ. However, we shall see that this simple *proportionality* does not apply to *all* events but only to the particular events described above.

We now have the correspondence $(c\beta\tau, v\beta\tau)_1 \leftrightarrow (c\tau, 0)_2$, which may be written $c\tau(\beta, \beta v/c)_1 \leftrightarrow c\tau(1, 0)_2$. Using proposition 1, we obtain for one of the unit vectors from P_2:

$$(14) \qquad (1, 0)_2 \leftrightarrow \left(\beta, \frac{\beta v}{c}\right)_1.$$

We wish to find a similar correspondence for the second unit vector $(0, 1)_2$. Suppose that

$$(15) \qquad (0, 1)_2 \leftrightarrow (a, b)_1,$$

and apply the fundamental law (11) to this event and that described by (14). We obtain

$$0 = -\beta a + \beta\frac{v}{c}b.$$

Then apply Eq. (12) to the corresponding vectors (15) to obtain

$$1 = -a^2 + b^2.$$

We now solve the last two displayed equations for a, b to obtain $b = \pm\beta$, $a = \pm(v/c)\beta$. We choose the solution $b = \beta$, $a = (v/c)\beta$ (why?), and (15) is now

(16)
$$(0, 1)_2 \longleftrightarrow \left(\beta\frac{v}{c}, \beta\right)_1.$$

We can now use (14) and (16) to obtain the event vector in P_2 corresponding to any given event vector (ct, x) in P_1. Suppose that $(ct, x)_1 \longleftrightarrow (c\bar{t}, \bar{x})_2$ and write the latter vector in terms of unit vectors:

$$(c\bar{t}, \bar{x})_2 = c\bar{t}(1, 0)_2 + \bar{x}(0, 1)_2.$$

Now use (14), (16), and proposition 1 to obtain

$$c\bar{t}(1, 0)_2 \longleftrightarrow c\bar{t}\left(\beta, \beta\frac{v}{c}\right)_1 \quad \text{and} \quad \bar{x}(0, 1)_2 \longleftrightarrow \bar{x}\left(\beta\frac{v}{c}, \beta\right)_1.$$

Then proposition 2 yields

$$(c\bar{t}, \bar{x})_2 = c\bar{t}(1, 0)_2 + \bar{x}(0, 1)_2 \longleftrightarrow c\bar{t}\left(\beta, \beta\frac{v}{c}\right)_1 + \bar{x}\left(\beta\frac{v}{c}, \beta\right)_1.$$

Since we also have $(c\bar{t}, \bar{x})_2 \longleftrightarrow (ct, x)_1$, we may write

$$(ct, x)_1 = c\bar{t}\left(\beta, \beta\frac{v}{c}\right)_1 + \bar{x}\left(\beta\frac{v}{c}, \beta\right)_1$$

$$= \left(\beta\frac{v}{c}\bar{x} + \beta c\bar{t}, \beta\bar{x} + \beta v\bar{t}\right)_1,$$

whence, equating the components, we obtain

(17)
$$x = \beta\bar{x} + \beta v\bar{t}$$
$$t = \beta\frac{v}{c^2}\bar{x} + \beta\bar{t}$$

These are the formulas we set out to obtain, and they constitute the *Lorentz transformation* from the P_2 coordinates $\bar{x}, \bar{t}$ to the P_1 coordinates x, t. The assumptions we have made, and particularly Einstein's law, imply that coordinates of the *same* event observed from P_1 and P_2 must be related in this way. The reader is urged to attempt the Exercises for Section 6.9 at this point in order to fix these ideas.

We have noted that times ascribed to the same event may differ. The same applies to the measures of distance between two fixed points. Thus, the relation (16) indicates that the distance $\bar{x}$ measured in P_2 at time $\bar{t} = 0$ will be observed as a distance $x = \beta\bar{x}$ if measured in P_1 at the same instant.

It should also be noted that if $v = 0$, then $\beta = 1$ (see definition (13)). Thus, if there is *no* relative velocity between P_1 and P_2, then these relativistic variations in time and distance measurements disappear, and the formulas reduce to the familiar classical situation which is more compatible with our physical intuition (see Exercise 2 of Section 6.2).

Exercises for Section 6.9

1. Sketch graphs of the following equations:

$$y = \sqrt{1 - x^2} \qquad (-1 \le x \le 1)$$
$$y = (\sqrt{1 - x^2})^{-1} \qquad (-1 < x < 1).$$

(*Hint:* Use the first to help with the second.)

2. Using the definition of β in Eq. (13), sketch a graph of β as a function of v for values of v between $-c$ and c.

3. Prove that the definition (13) implies $v = c\sqrt{1 - \beta^2}$.

4. Suppose that a particle has velocity, u, observed from a platform, P_1, and v observed from platform P_2. If it is found that $|u| = c$, the speed of light, show that $|v| = c$ also.

5. Prove that the Lorentz transformation in the form of Eqs. (17) implies

$$\bar{x} = \beta x - \beta v t$$
$$\bar{t} = -\beta \frac{v}{c^2} x + \beta t$$

Show that these equations are consistent with classical theory at speeds $|v|$ very much less than c (see Exercise 2 of Section 6.2).

6. (a) A rigid rod lies at rest in a platform P_1 and the coordinates of its end points are x_0 and $x_0 + a$. What will the length of the rod be as observed from P_2? (*Hint:* Use the first of the Eqs. (17) and note that the ends of the rod are observed at the same instant as measured in P_2. The rod should appear to be shrunk by a factor $1/\beta$ when observed from P_2.) Similarly, show that a rod fixed in P_2 appears shrunk when observed from P_1.

    (b) Estimate the apparent difference in length in the readings of part (a) if $a = 1$ meter and $v = 1000$ meters/sec. What if $v = c/3$?

7. After completing Exercise 6, show that the relativity theory confirms the experimental result of the Michelson-Morley experiment.

6.10   Velocity, Momentum, and Energy

Since the measurements of time and distance require serious reexamination as we go from classical to relativistic theories, it should not be surprising that the concepts of velocity, momentum, and energy also need some reexamination. Our descriptions of momentum and energy will follow fairly naturally when we have solved the problem of the appropriate description of velocity.

Suppose that a particle travels with velocity, v, through a platform P_1. We would like to assign a *vector* to this velocity, and we would like this velocity vector to be independent of the platform of observation, in an appropriate sense. If the particle passes the origin of P_1 at time $t = 0$, then at time t it is at a distance vt from the origin. Thus the event vector giving its location at any time t is $(ct, vt)_1$. If we simply divide by t (multiply by $1/t$), we obtain a "velocity-like" vector $(c, v)_1$. The dependence on the platform is, however, very clear. Our best hope of obtaining a vector independent of the platform is now to find the event vector corresponding to $(c, v)_1$ in a platform P_2 *moving with the particle*. Multiplying (14) by c/β, we find that

$$(c, v)_1 \longleftrightarrow (c/\beta, 0)_2.$$

But this still depends on P_1 because a β (depending on v) appears in the first component.

However, using proposition (1) of Section 6.8, we can multiply both sides of the above correspondence by β to obtain $(\beta c, \beta v)_1 \longleftrightarrow (c, 0)_2$. Since the latter vector *is* independent of v, we choose $(\beta c, \beta v)_1$ for the *relativistic velocity vector* as observed from P_1.

It is important that we see the logic behind this choice. If we observe the uniform motion of a particle from platforms having different velocities, we obviously obtain different vector representations of the velocity of the particle. The beauty of our definition is that, whatever platform we use to formulate the velocity vector, we obtain precisely the same vector $(c, 0)$ on calculating the corresponding vector for the platform moving with the particle. This consistency is necessary if the laws of mechanics which involve the velocity are to be independent of the platform of observation (cf. Section 6.4).

It is now plausible to try and define a momentum vector for the particle by

$$p = m(\beta c, \beta v)_1 = (m\beta c, m\beta v)_1,$$

where m is the mass of the particle. Thus, the momentum vector is just the mass multiplied by the velocity vector. The second component of this vector (called the *space* component) is $m\beta v$, and when v/c is very small, then β is almost 1 and this component is nearly mv, the momentum as defined in classical mechanics. We now describe $m\beta$ as the *apparent mass* which reduces to the *rest mass* m when $v = 0$.

Consider now the first (or *time*) component of the vector p. If we write $p = (p^{(0)}, p^{(1)}) = (m\beta c, m\beta v)$, then we have $p^{(0)} = m\beta c$ and

$$cp^{(0)} = mc^2\beta = mc^2 \left(1 - \frac{v^2}{c^2}\right)^{-1/2}.$$

If v/c is very small,* we have

(18) $$cp^{(0)} \simeq mc^2 + \tfrac{1}{2}mv^2.$$

But this now looks like an energy! We know that $\tfrac{1}{2}mv^2$ is the kinetic energy of the particle with its rest mass. We therefore describe the quantity

(19) $$cp^{(0)} - mc^2 = mc^2(\beta - 1)$$

as the kinetic energy of the particle, and the definition implies that if v/c is very small, the energy of the particle is approximately $\tfrac{1}{2}mv^2$, as it should be. We are also tempted to describe mc^2 as some form of energy, but we postpone this discussion for the moment. Nevertheless, let us agree to call mc^2 the *rest energy* of the particle.

We see now that the vector p contains information on both energy and momentum, and it is therefore described as the *energy-momentum vector* of the particle.

Exercises for Section 6.10

1. (a) What is greater, rest mass or apparent mass?
   (b) A particle has mass m and velocity $(0.6)c$ measured in the platform P_1. Find its apparent mass, relativistic velocity vector, energy-momentum vector, (relativistic) kinetic energy, and rest energy.
2. Sketch graphs of (a) the space component of the energy-momentum vector and (b) the kinetic energy of a particle as functions of v for $-c < v < c$. (Use Exercise 2 of Section 6.9.)

6.11 Collisions Revisited

Having discussed momentum and energy of a particle as described in relativistic terms, we now have all the necessary tools to investigate collision processes in the same theory. Let us first decide what we expect of such a theory. The governing laws should be independent of the choice of platform of observation and should, of course, be based on the velocity and energy-momentum concepts just introduced. When all the velocities in the process are small compared to c, the new models should reduce to corresponding models constructed in Sections 6.3, 6.5, and 6.6 for collisions in the classical theory. The first situation we consider is that in which particles with rest masses m_1, m_2 have initial velocities u_1, u_2. After collision, the rest masses are the same and the velocities are v_1, v_2, respectively. To emphasize the

*See Appendix 1 of Chapter 3.

dependence of β on the velocity, we write β_v for β in the definition (13). For the energy-momentum vectors we have:

Before impact: $\quad p_1 = m_1(\beta_{u_1}c,\ \beta_{u_1}u_1), \quad p_2 = m_2(\beta_{u_2}c,\ \beta_{u_2}u_2).$

After impact: $\quad q_1 = m_1(\beta_{v_1}c,\ \beta_{v_1}v_1), \quad q_2 = m_2(\beta_{v_2}c,\ \beta_{v_2}v_2).$

It is tempting to investigate the properties of a model in which the total energy-momentum *vector* is conserved. Thus, we take as our first postulate

$$(20) \qquad\qquad p_1 + p_2 = q_1 + q_2.$$

Since this is an equation between vectors, we obtain two equations in real numbers by comparing the two components. Consider first the time components which give:

$$(21) \qquad\qquad m_1\beta_{u_1} + m_2\beta_{u_2} = m_1\beta_{v_1} + m_2\beta_{v_2}.$$

This says merely that the total apparent mass is conserved.

When all speeds concerned are small compared to c and we multiply both sides of the last equation by c^2, we obtain from (18)

$$(m_1c^2 + m_2c^2) + \tfrac{1}{2}m_1u_1^2 + \tfrac{1}{2}m_2u_2^2 \simeq (m_1c^2 + m_2c^2) + \tfrac{1}{2}m_1v_1^2 + \tfrac{1}{2}m_2v_2^2.$$

Thus we retrieve the kinetic energy equation for *perfectly elastic* impact as discussed in Section 6.3.

The space component of our conservation law yields

$$m_1\beta_{u_1}u_1 + m_2\beta_{u_2}u_2 = m_1\beta_{v_1}v_1 + m_2\beta_{v_2}v_2.$$

This is rather like the momentum law used in Section 5.3 except that the mass is replaced by the apparent mass. Furthermore, since β is almost 1 if $|v|$ is small compared to c, the equation does reduce to the classical conservation of momentum in this case.

We interpret these results in the following way: In the relativistic formulation, perfectly elastic impact is characterized by the fact that the rest masses of the particles remain unchanged. With this hypothesis, the conservation of the energy-momentum vector *implies* that the model reduces to that of perfectly elastic impact when the speeds are small compared to c.

We now have a model of elastic impact built up entirely by mathematical argument based on certain physical prerequisites of the model. The argument is, in one respect, tidier than the classical formulation of the problem, since we require only one conservation law—although this admittedly involves two components of a single law concerning vectors. Are there real-life situations for which this model is appropriate? We must simply accept the fact that there are, but these situations are outside our everyday experience. Some important situations where the new model is needed are in the collisions of the elementary particles of modern physics. In particle accelerators, electron speeds within 2% of the speed of light are common. In these circumstances, the relativity theory is in agreement with experiment and the classical theory is not.

Before we consider inelastic collisions, it should be carefully noted that the elastic case can be arrived at if we first assume the conservation law and then say something about the behavior of the rest masses. This suggests that, for inelastic impact in which the particles combine, we might first postulate the *same* conservation law and then examine the way in which the masses combine.

Suppose, then, that particles with rest masses m_1, m_2 and initial velocities u_1, u_2, respectively, collide and combine into a particle of mass m_0. We can simplify the analysis by choosing as our platform of observation one which moves with the combined mass (see Section 6.4). After the impact, the velocity vector for m_0 will therefore be $(c, 0)$. We now have:

Before impact: $\quad p_1 = m_1(\beta_{u_1}c, \beta_{u_1}u_1), \quad p_2 = m_2(\beta_{u_2}c, \beta_{u_2}u_2).$

After impact: $\quad q = m_0(c, 0).$

The time component of the conservation law now reads

(22) $$m_1\beta_{u_1} + m_2\beta_{u_2} = m_0.$$

Now if $u_1 \neq 0$, $u_2 \neq 0$, then $\beta_{u_1} > 1$ and $\beta_{u_2} > 1$, and we deduce that $m_0 > m_1 + m_2$. Thus it seems that the resultant rest mass actually exceeds the sum of the two initial rest masses. This implies the same relation between the rest energies:

$$m_0 c^2 > m_1 c^2 + m_2 c^2.$$

Furthermore, we easily deduce from Eq. (22) that

$$(m_1\beta_{u_1}c^2 - m_1 c^2) + (m_2\beta_{u_2}c^2 - m_2 c^2) = m_0 c^2 - m_1 c^2 - m_2 c^2.$$

Since the expressions on the left are (by (19)) merely the kinetic energies of the particles before impact, and the energy of the particle after impact is zero, we state the following rule based on conservation of the energy-momentum vector: The loss in kinetic energy in an inelastic impact is equal to the increase in the rest energy.

Compare this result with that of the classical model of perfectly inelastic impact. In that case, we also observed a loss in kinetic energy, but the model gave us no clue as to the way in which this energy may be absorbed. The relativistic model suggests that the kinetic energy lost is balanced by an increase in the rest energy.

It was a fundamental concept of Einstein that the rest energy of a particle, mc^2, may be viewed as a stored or internal energy. The acceptance of this idea allows us to present an appealing, self-contained model (for appropriate physical situations) which does not require some unknown mechanism for the absorption of lost kinetic energy, as discussed in Section 6.5. However, when we come to investigate the reverse, or explosion process, we are immediately faced with the idea that this rest energy, mc^2, may, under appropriate conditions, be available in that it can be transformed into kinetic energy.

In an explosion involving the decomposition of a molecule into atoms, the internal energy available as rest energy arises from the chemical bonds which initially bind the atoms together. Nuclear energy, however, arises on breaking the much stronger nuclear bonds which bind together the fundamental particles making up the nucleus of an atom.

Note, finally, that the energy released in an explosion process is, according to this theory, dependent on a reduction in mass, and we are left with the intriguing possibility that the rest energy of a particle may be completely transformed into the kinetic energy of particles with zero mass! It turns out that even this is not too wide a stretch of the imagination, and such possibilities, predicted on a theoretical basis, have indeed been observed in nature.

Exercises for Section 6.11

1. A particle of mass 1 strikes a particle of mass 10 which is at rest, with a velocity of $(0.8)c$. Show that the perfectly elastic relativistic model predicts that the resulting speeds v_1 and v_2 satisfy

$$\beta_{v_1} + 10\beta_{v_2} = 11\tfrac{2}{3}$$

$$\beta_{v_1}v_1 + 10\beta_{v_2}v_2 = \tfrac{4}{3}c.$$

2. A perfectly inelastic collision is observed from a platform from which the resulting single particle is observed to be at rest. If, before impact, the particle of mass m_1 had velocity $\tfrac{4}{5}c$ and the particle m_2 had velocity $-\tfrac{3}{5}c$, show that:
   (a) The resulting particle has mass $\tfrac{5}{3}m_1 + \tfrac{5}{4}m_2$.
   (b) The rest energy gained in the impact is $\tfrac{2}{3}m_1c^2 + \tfrac{1}{4}m_2c^2$.
   (c) $\dfrac{m_1}{m_2} = \dfrac{9}{16}$.

3. A perfectly inelastic collision occurs between two particles, one of 6 units of mass and the other of 1 unit of mass. If their velocities before impact are $(0.8)c$ and $(0.6)c$, respectively, show that the resulting particle has speed $7c/9$ and $5\sqrt{2}$ units of mass. Compare these results with those predicted by a Newtonian model of the same process.

4. A particle of rest mass m_0 at rest in the frame of observation decomposes explosively into two particles whose velocities are observed to be $3c/5$ and $7c/25$ in opposite directions. Find the rest masses of the two particles and find the loss in rest energy.

Appendix I: A Summary of Some Models Discussed in Chapter Six

	Newtonian	Relativistic
Perfectly elastic collision	1. Masses are unchanged. 2. Momentum is conserved. 3. Kinetic energy is conserved.	1. Masses are unchanged. 2. Energy-momentum vector is conserved.
Perfectly inelastic collision	1. Particles combine. 2. Mass is conserved. 3. Momentum is conserved.	1. Particles combine. 2. Energy-momentum vector is conserved.
Explosion	1. Particle separates. 2. Mass is conserved. 3. Momentum is conserved. 4. Energy law.	1. Particle separates. 2. Energy-momentum vector is conserved.

Appendix II: Corresponding Event Vectors

If the same event is described from two different platforms we write, as on p. 146,

$$(ct, x)_1 \longleftrightarrow (c\bar{t}, \bar{x})_2.$$

Thus, t, x give the time and position of the event as observed from the first platform, and $\bar{t}$, $\bar{x}$ give the time and position of the *same* event observed from the second platform. We wish to prove part 1 of the proposition at the end of Section 6.8. Part 2 is relatively easy to prove and is given as an exercise.

Proof. We suppose that

(A.1) $$k(ct, x)_1 \longleftrightarrow (c\tau, \xi)_2,$$

and we must prove that $\tau = k\bar{t}$, $\xi = k\bar{x}$.

Consider first the event described by $(1, 1)_1$ and let $(1, 1)_1 \longleftrightarrow (\alpha, \beta)_2$. Then applying the fundamental law in the form of Eq. (12) to this vector, we deduce that

$$-1^2 + 1^2 = -\alpha^2 + \beta^2,$$

or $\beta^2 = \alpha^2$, which implies $\beta = \pm\alpha$. Since we have assumed that $(0, 0)_1 \leftrightarrow (0, 0)_2$, we may write

(A.2) $(1, 1)_1 \leftrightarrow (\alpha, \pm\alpha)_2$ $(\alpha \neq 0)$.

We now apply Einstein's basic law (11) to the events described by $(ct, x)_1 \leftrightarrow (c\bar{t}, \bar{x})_2$ and by (A.2), and then to the events (A.1) and (A.2). We obtain

(A.3) $-ct + x = -\alpha(c\bar{t} \pm \bar{x})$

(A.4) $k(-ct + x) = -\alpha(c\tau \pm \xi)$.

Substituting for $-ct + x$ from (A.3) into (A.4) and using the fact that $\alpha \neq 0$, we obtain

(A.5) $c\tau \pm \xi = k(c\bar{t} \pm \bar{x})$.

We then apply the basic law once more, this time to the events $(ct, x)_1$ and that of (A.1):

$$k(-c^2t^2 + x^2) = -c^2\bar{t}\tau + \bar{x}\xi$$

or, since $-c^2t^2 + x^2 = -c^2\bar{t}^2 + \bar{x}^2$,

$$c^2\bar{t}\tau - \bar{x}\xi = k(c^2\bar{t}^2 - \bar{x}^2).$$

Solving this equation simultaneously with (A.5) for τ and ξ, we find that, if $|\bar{x}| \neq |c\bar{t}|$, then $\tau = k\bar{t}$ and $\xi = k\bar{x}$ as required.

This is not the complete result, of course, and we still have to complete the proof for events vectors $(ct, x)_1$ with $|ct| = |x|$, i.e., vectors of the form $(\alpha, \pm\alpha)_1$. We have, however, proved enough to establish Eq. (14):

(A.6) $\left(\beta, \beta\dfrac{v}{c}\right)_1 \leftrightarrow (1, 0)_2$.

If we now suppose that

$$(\alpha, \pm\alpha)_1 \leftrightarrow (\gamma, \pm\gamma)_2$$

and

$$k(\alpha, \pm\alpha)_1 \leftrightarrow (\delta, \pm\delta)_2,$$

then combining each of these with (A.6) by means of the basic law, we obtain

$$\alpha\beta\left(-1 \pm \dfrac{v}{c}\right) = \gamma, \qquad \alpha\beta\left(-1 \pm \dfrac{v}{c}\right) = \delta,$$

whence $\delta = k\gamma$. Thus $(\alpha, \pm\alpha)_1 \longleftrightarrow (\gamma, \pm\gamma)_2$ implies that $k(\alpha, \pm\alpha)_1 \longleftrightarrow k(\gamma, \pm\gamma)_2$, and the proof is now complete.

Exercise: Prove part 2 of the proposition.
   Hints: Suppose that

$$(ct_1 + ct_2, x_1 + x_2)_1 \longleftrightarrow (c\tau, \xi)_2$$

and prove that $\tau = \bar{t}_1 + \bar{t}_2$ and $\xi = \bar{x}_1 + \bar{x}_2$. Use Eqs. (14) and (16) to do this. This is legitimate since only proposition 1 is used in their derivation, and this is already proved.

Preliminary Exercises

Complete Preliminary Exercises 1–4 of Chapter 2, and 8–13 of Chapter 3, before embarking on this set.
1. Solve for x in terms of a and b:

(a) $ax + b^2 = 3 - 2x$

(b) $\dfrac{a^2 + b^2}{x} = x + \dfrac{2ab}{x}$

2. (a) Solve for x and y in terms of a:

$$x - 3y = 2a$$
$$2x + y = 5a.$$

(b) Solve for u and v in terms of a and b, assuming $b \neq a$:

$$au + bv = a^2 + b^2$$
$$u + v = 2a.$$

Describe the solution set if $a = b$.
3. Solve the following pair of equations by elimination:

$$x + y = 7$$
$$x^2 + y^2 = 25.$$

Sketch graphs of the solution sets of the separate equations to confirm your answer.
4. Solve for x and y in terms of a and b, given that $|b| > |a|$:

$$x + ay = 2b, \qquad xy = a.$$

5. Let a, b, c, d satisfy $ad - bc = 1$. Solve

$$ax + by = u$$
$$cx + dy = v$$

for x, y in terms of u and v to obtain

$$du - bv = x,$$
$$-cu + av = y.$$

What is the result of solving *these* equations for u, v in terms of x and y?

BIBLIOGRAPHY

General References

Mathematics in the Modern World, Readings from *Scientific American*, W. H. Freeman and Co., San Francisco, 1968.

The Mathematical Sciences, A collection of essays, M.I.T. Press, Cambridge, Mass., and London, 1969.

"The Future of Applied Mathematics," *Quarterly of Applied Mathematics*, Vol. 30, (1972), No. 1.

J. G. KEMENY and J. L. SNELL, *Mathematical Models in the Social Sciences*, M.I.T. Press, Cambridge, Mass., and London, 1972.

M. S. KLAMKIN, "On the Ideal Role of an Industrial Mathematician and its Educational Implications," *American Mathematical Monthly* Vol. 78 (1971), pp. 53–76.

Chapter One

J. L. SYNGE and B. A. GRIFFITH, *Principles of Mechanics*, McGraw-Hill Book Company, New York, 1959. (Especially Chapter I.)

J. G. KEMENY and J. L. SNELL, *Mathematical Models in the Social Sciences*, M.I.T. Press, Cambridge, Mass., and London, 1972.

Chapter Two

A. M. GLICKSMAN, *Linear Programming and the Theory of Games*, John Wiley & Sons, Inc., 1963.

S. LIPSCHUTZ, *Finite Mathematics*, Schaum Publishing Co., New York, 1966.

P. M. COHN, *Linear Equations*, Routledge and Kegan Paul, London, 1958.

B. C. TETRA, *Basic Linear Algebra*, Harper & Row, Publishers, New York, 1969.

Chapter Three

N. KEYFITZ, *Introduction to the Mathematics of Population*, Addison Wesley Publishing Company, Inc., 1968.

D. P. ODUM, *Fundamentals of Ecology*, W. B. Saunders, Philadelphia, London, Toronto, 1971.

J. MAYNARD SMITH, *Mathematical Ideas in Biology*, Cambridge University Press, 1968.

E. O. WILSON and W. H. BOSSERT, *A Primer of Population Biology*, Sinauer Associates, Stanford, U.S.A., 1971.

L. R. CLARK, P. W. GEIER, R. D. HUGHES, and R. F. MORRIS, *The Ecology of Insect Populations in Theory and Practice*, Methuen, London, 1967.

I. A. McLAREN, *Natural Regulation of Animal Populations*, Atherton Press, New York, 1971.

Chapter Four

L. COOPER and D. STEINBERG, *Introduction to Methods of Optimization*, W. B. Saunders, Co., Philadelphia, 1970.

D. J. WYLDE, *Optimum Seeking Methods*, Prentice-Hall, Inc., Englewood Cliffs, N.J., 1964.

Chapter Five

S. GOLDBERG, *Probability, An Introduction*, Prentice-Hall, Inc., Englewood Cliffs, N.J., 1960.

C. DERMAN, L. J. GLASER, and J. OLKIN, *A Guide to Probability Theory and Application*, Holt, Rinehart & Winston, Inc., New York, 1973.

C. C. LI, *Population Genetics*, U. of Chicago Press, 1955.

W. D. STANSFIELD, *Genetics*, Schaum's Outline Series, McGraw-Hill Book Company, New York, 1969.

Chapter Six

K. O. FRIEDRICHS, *From Pythagoras to Einstein*, L. W. Singer Co., 1965.

B. RUSSELL, *The ABC of Relativity*, Harper Bros., 1925.

J. L. SYNGE and B. A. GRIFFITH, *Principles of Mechanics*, McGraw-Hill Book Company, New York, 1959.

R. DUGAS, *A History of Mechanics*, Editions du Griffon, Neuchatel, 1955.

INDEX

Absolute value, 48, 75
Applied mathematics, 1, 5, 126

Binomial expansion, 54, 74
Bisexual population, 112

Catastrophe, 61-63
Clock, 127, 128
Collisions, 126-157
   inelastic, 136-139, 155-157
   perfectly elastic, 130-136, 154, 156, 157
Competition, of species, 64, 65, 72
Constraint, 12, 13, 18, 24
Contour, 20, 23
Control theory, 73

Density, of population, 42-44
Diminishing returns, 79, 80
Discrete generations, 112
Distance, 48, 75
Dominant genotype, 111

Einstein, 144, 145, 156
Einstein's law, 146, 147, 150, 158
Elementary operations, 27
Elimination method, 26, 159
Energy, conservation of, 132, 135, 137, 139
   internal, 137, 140, 156
   kinetic, 132, 137, 153, 155-157
Equilibrium, 38, 41-43, 52-55, 116, 124
   stable, 41-44, 56, 117
   unstable, 41, 42, 46
Equivalent equations, 27
Error analysis, 48-51
Event:
   in probability theory, 106
   in relativity theory, 145
   origin, 146

Events, simultaneous, 145
Evolution, 73, 104, 114
Experiment:
   in search techniques, 78
   in probability theory, 105
Exploitation, 65
Explosions, 139-141, 156
Exponent laws, 75, 103
Extinction, of species, 41

Faraday, 143
Feasible solution, 13, 18, 19, 24
Fibonacci numbers, 101
Flat-earth theory, 6
Force, 127, 129
Frame of reference, 128, 129
Function, 41
   linear, 15
   differentiable, 80
   unimodal, 80, 81, 83

Gene, 110
   types, 111
Gene frequency, 121, 122
Genotype, 111
Geocentric theory, 6
Geometric progression (series), 50, 52, 67, 79, 92

Half-plane:
   closed, 16, 18
   open, 16
Hardy-Weinberg law, 116, 117, 121-123
Hertz, 143
Host-parasite interaction, 64, 66
Hybrid genotype, 111

Idealization, 2, 8, 10, 42, 99, 127, 131
Inbreeding, 119